Books by the same author:

- *Journey through the World of Malaria*
- *Ebola Virus: A Natural Killing Machine: A Critical Review*
- *HIV/AIDS: Recent Advances*
- *Man: The Wonderful Creature of God: Scientific & Religious Evidence*

ONCOVIRUSES: CELLULAR

AND

MOLECULAR VIROLOGY

Abubakar Yaro (Khalifa)

authorHOUSE®

AuthorHouse™ UK
1663 Liberty Drive
Bloomington, IN 47403 USA
www.authorhouse.co.uk
Phone: UK TFN: 0800 0148641 (Toll Free inside the UK)
         UK Local: 02036 956322 (+44 20 3695 6322 from outside the UK)

Published by AuthorHouse 04/21/2021

ISBN: 978-1-6655-8854-6 (sc)
ISBN: 978-1-6655-8853-9 (hc)
ISBN: 978-1-6655-8855-3 (e)

Print information available on the last page.

CONTENTS

PREFACE

Oncoviruses are cancer-causing viruses that play a leading role in the development of certain cancers by contributing to genetic changes that result in the disruption of cell cycle machinery, interfering with certain key functions such as cell growth. Human viral oncogenesis has common traits:

1. Oncoviruses are required but not sufficient for cancer development.
2. Virus-associated cancers appear in the context of persistent infection and occur up to a decade after acute infection.
3. The immune system can have a protective or damaging effect, with some human virus-associated cancer increasing with immunosuppression and others appearing in the context of chronic inflammation.

Oncoviruses: Cellular and Molecular Virology presents a state-of-the-art review based on original research covering some interesting aspects of oncoviral infections and the oncogenetic process. The book is made of ten critical chapters. It begins with a discussion of the molecular epidemiology of human oncoviruses. It explores several associated topics, including oncoviruses and the immune system; envelope antigens of oncoviruses; oncoviral integration, with Epstein-Barr virus (EBV) as a model; the evolution and pathogenesis of HTLV; the P53 antigen; HPV oncoprotein; viral-host interaction in HBV-associated HCC; Kaposi's sarcoma-associated herpesvirus; and utilization of flow cytometry in the development of therapeutic agents. This book will be of interest to virologists, oncologists, public health specialists, and palliative care providers.

ACKNOWLEDGEMENTS

My greatest gratitude to my Creator for giving me the required knowledge and aptitude, which empowered me to undertake this ambitious project. My next thanks go to my parents (Abdullahi Yaro and Hajia Amina Ahmed), who tasked me to do all I could to become educated. It was this challenge that helped me to stay focused during the most difficult part of my life. To my late sisters, Maryam and Abiba, your love for me was and is the guiding star in my life. I thank you for being great siblings to me.

To all individuals associated with Africa Health Research Organization, I thank you immensely for wonderful commitment to a cause.

During my period of studies, I met a lot of lecturers who taught me great things. To all of you, accept my sincere and humble gratitude for the energies you expended in turning me from an angry man into a scientist. One man who kick-started my writing career is Professor Popejoy. Pope, thanks so much.

My children became my bed of comfort. They gave me hope that I will leave a dynasty when I am gone. Hi, kids, thanks for your natural love.

Winsome Morgan, a very brilliant and hard-working nurse, was very instrumental in my life. She made many financial contributions to help stabilise my academic career. What I am today and where I have reached was partly due to her great financial commitment. Winsome, thank you so much.

My greatest scientific thanks goes to Professor Pranab K of B Medical College India, who I regard as one of the greatest pathologists of our generation, for devoting so much time reviewing this book. His valuable contribution turned this book into an interesting write-up.

To my wives: Hajia Ayisha and Hajia Maimuna, you are the guardians of the Abubakar dynasty. Thanks so much for your love, which keeps my ambition alive.

DEDICATION

This book is dedicated to my parents, Abdullahi Yaro and Hajia Amina Ahmed, my late sisters, Maryam and Abiba, my children, and Winsome Morgan.

CHAPTER 1

MOLECULAR EPIDEMIOLOGY OF HUMAN CANCER VIRUSES

1.1 Background

About 12 percent of human cancers globally are caused by oncoviral infection, with more than 80 percent of these found in the developing countries. Human viral-associated cancers are of public health importance and suitable for immunoprophylaxis and targeted therapies. However, there are still formidable challenges to understanding and managing viral-associated cancers: limitation in animal models of the disease, the disparate nature of these cancers, the different types of viruses that caused them, and the complex nature of the virus-host interaction, which leads to the development of cancer.

Oncoviruses can be either DNA or RNA viruses (table 1). DNA tumour viruses have a DNA genome that is transcribed into RNA, which is then translated into proteins. These viruses have two life cycles: permissive and non-permissive. In permissive cells, all parts of the viral genome are expressed. Expression of the genome leads to virus replication, cell lysis, and cell death. In cells that are non-permissive for replication, viral DNA is usually (but not always) integrated into the cell chromosome at random sites. Only part of the viral genome is expressed. This is the early, control functions of the virus. Viral structural proteins are not manufactured, and no progeny viruses are released.

The RNA tumour differs from the DNA tumour in that the genome is RNA but is similar to DNA tumour viruses, in having the genome integrated into the host genome. Since RNA makes up the genome of the mature particles, it must be copied to DNA before it is integrated into the host cell chromosome. This strategy runs contrary to the central dogma of molecular biology, in which the DNA is copied into RNA.

DNA tumour viruses have been implicated in the aetiology of human cancers, including human papillomavirus (HPVs), Epstein-Barr virus (EBV), Kaposi's sarcoma (KS)-associated herpesvirus (KSHV), hepatitis B virus (HBV), and Merkel cell polyoma virus (MCV). Among the RNA viruses, hepatitis C virus (HCV) and human T-cell leukaemia virus type-1 (HTLV-1) are associated with human cancers. Infection with human immunodeficiency virus (HIV) is associated with cancer incidence, although immunodeficiency may be a contributing factor. According to the International Agency for Research on Cancer (IARC), the lead cause of cervical cancer is infection with HPV 16, 18, 31, 33, 35, 45, 51, 52, 56, 58, 89, and 66.

EBV was first identified in 1964 and was the first recognised herpesvirus to be identified as oncogenic in humans. EBV is associated with four types of malignancies: Burkitt's lymphoma, Hodgkin's lymphoma, nasopharyngeal carcinoma, and non-Hodgkin's lymphoma (NHL) linked with post-transplant or immunosuppression by HIV. KSHV is also referred to as human herpesvirus 4 (HHV-4) and as HHV-8. In the early 1990s, it was reported that Kaposi's sarcoma was one of the apparent clinical manifestations of acquired immunodeficiency syndrome (AIDS). KSHV was subsequently reported from a case of AIDS-associated KS in 1994.

HBV is a single-stranded RNA virus that is transmitted mostly via contaminated blood or unsafe medical procedure. The World Health Organization (WHO) estimates that about 170 million people are infected worldwide, although the number keeps increasing at an alarming rate. HBV is one of the causes of liver cancer (hepatocellular carcinoma, HCC), the sixth most common cause of cancer in the world, and the third in terms of mortality. It is estimated that 54 percent of the global cancer burden can be attributed to HBV, and 31 percent to HCV infections. The development of HCC is generally slow, over a period of more than thirty years after infection with HBV or HCV.

HTLV-1 was the first reported retrovirus isolated in 1980 and 1981 from American and Japanese patients suffering from adult T-cell leukaemia (ATL). The causal relationship between the virus and malignancy has been established. It is mostly transmitted in three ways: mother to child, through breastfeeding; sexually via partners; or iatrogenically through transfusion of blood products. It is estimated that 15–20 million people are chronic HTLV-1 carriers, up to 5 percent of whom are at risk of developing ATL.

A human T-cell leukaemia type II (HTLV-II) has been described. Unlike HTLV-I, it shows tropism for CD4+ lymphocytes, preferentially infecting CD8+ lymphocytes. HTLV-II is endemic in a number of geographic regions, such as North Africa and parts of North and South America. Despite its prevalence, there is limited evidence to link it to human diseases. Most cases of HTLV-II infection are associated with haematologic disturbances such as pancytopenia, atypical hairy cell leukaemia, lymphatic leukaemia, large cell lymphoma, and mycosis fungoids. HTLV-II infection may also produce neurological disorders such as cognitive dysfunction. With increasing incidence and prevalence of viral-associated cancers, it is important to identify more research theme on HTLV-II.

Historically, on 1 October 1909, Francis Peyton Rous started searching for oncoviruses when he began studying cancer-virus transmission at Rockefeller University in the United States. Rous successfully transplanted a sarcomatous chest tumour from a fifteen-month-old hen into other chickens. By 1911, he had successfully showed the cancer could be transmitted via cell-free tumour extracts. He concluded that it might be a virus that caused that tumour. Earlier in 1908, two Danish scientists, Oluf Bang and Vilhelm Ellerman, had published a paper on viral transmission of avian erythoblastosis. Rous gave up his studies on viral cancers until the 1930s, when mammalian tumour cancer was described. In 1934, with his colleague Richard Shope, Rous returned to viral tumour biology using cottontail rabbit papillomavirus. In the 1950s, interest in viruses as cause of cancer grew after Ludwick Gross discovered an acute transforming murine retrovirus. During the same period, mouse leukaemia virus and polyoma virus were discovered.

In 1966,Rous was awarded the Nobel Prize for discovering Rous sarcoma virus. In the 1960s, EBV tumour virus was first discovered in humans. In the 1970s, epidemiological studies linked HBV infection to HCC. HPV was associated with cervical carcinoma in a proposal by zur Hausen in the 1970s. This discovery led to the development of Cervarix and Gardasil, two anticancer agents that protect against infection with HPV16 and HPV18, the cause of most cervical cancer. During the same decade, clustering cases of leukaemia in south-west Japan led to the isolation and description of a retrovirus which was shown to be identical to adult T-cell leukaemia.

In 1989, John Michael Bishop and Harold E. Varmus were awarded the Nobel Prize for discovering the viral oncogene c-Src. Advances in molecular techniques had positive impact in the field of oncovirology. Two new oncoviruses were discovered. In 1994, Chang et al., using a PCR-based technique, discovered Kaposi's sarcoma-associated herpesvirus (KSHV). Later, Merkel cell polyoma virus (MCV) was discovered. It is the only human polyoma virus with proven oncogenic ability among the many human polyoma viruses. MCV was discovered among patients with Merkel cell carcinoma using digital transcriptome subtraction (DTS). In 2008, the Nobel Prize was awarded to zur Hausen for discovering that high-risk HPV causes cervical cancer and to Luc Montagnier and Francois Barre-Sinoussi for the discovery of HIV, a virus that does not directly cause cancer but initiates the stage for cancer growth through immunosuppression.

Virus	Associated Cancers	Mode of Transformation	Mechanism of Carcinogenesis
HPV	Cancers of the cervix, anus, penis, vulva, vagina, oropharynx	Direct	Production of the proteins E6 and E7

HBV	HCC	Indirect	Chronic inflammation
EBV	Burkitt's lymphoma, Hodgkin's lymphoma, nasopharyngeal carcinoma, non-Hodgkin lymphoma associated with post-transplant or HIV immunosuppression	Direct	Production of viral oncoprotein during lytic infection
HCV	HCC	Indirect	Chronic inflammation
KSHV	Kaposi's sarcoma, PEL, MCD	Direct	Production of viral oncoprotein during lytic infection
MCV	MCC	Direct	Viral genes, large and small tumour antigen
HTLV-1	ATL	Direct	Production of viral oncoprotein Tax
HIV	Kaposi's sarcoma	Direct	Immunosuppression

Table 1: Characteristics of some oncoviruses

1.2 Prevalence of Some Oncoviral Infections

EBV is highly prevalent around the globe; It is estimated that around 5.5 billion are infected with EBV globally. Two major types of EBV have been identified, differing in geographical distribution. EBV-2 is more common in Africa and homosexual men. HBV infects more than 2 billion worldwide, and more than 300 million are chronic HBV carriers, with an estimated 1 million deaths annually from HBV-associated liver diseases, including severe complications such as liver cirrhosis and HCC. The prevalence is highest in sub-Saharan Africa, the Amazon basin, China, Taiwan, and several countries in South East Asia. In areas of high endemicity, the lifetime risk of HBV infection is more than 60 percent, with more infections acquired via perinatal and child-to-child transmission, where the risk of infection becoming chronic is greatest. Vertical transmission is predominant in China, Korea, and Taiwan, while in sub-Saharan Africa, child-to-child transmission is most common.

By comparison, HCV infects around 150 million people worldwide, with an estimated 2.2 per cent prevalence. The estimate of HCV ranges from <0.1per cent in the UK to 15–20 per cent in Egypt. HCV infection is highly prevalent in Mongolia, northern Africa, China, Pakistan, southern Italy, and some areas of Japan. Six major genotypes of HCV have been described: genotype 1, genotype 2, genotype 3, genotype 4, genotype 5, and genotype 6. About 75 per cent of Americans with the virus have genotype 1 (subtype 1a or 1b), and 20–25 per cent have genotype 2 or 3, with smaller numbers infected with genotypes 4, 5, or 6. Genotype 4 is more common in Africa, while genotype 6 is common in South East Asia. HCV has two major routes of transmission: IV drug use and iatrogenic exposure through transfusion, transplantation, or unsafe therapeutic interventions.

By the end of 2014, HIV infection was found in an estimated 36.9 million people globally. An estimated 0.8 per cent of adults aged 15–49 years are living with the virus worldwide. Sub-Saharan Africa continues to house the biggest burden, with 70 per cent in 2014. After sub-Saharan Africa, the Caribbean, eastern Europe, and central Asia are the most severely affected. There were 2 million, among them 390,000 children, who were newly infected with HIV at the end of 2014. HIV-1 is transmitted via three major routes: sexual intercourse, blood contact, and mother-to-child transmission.

Cervical cancer is the most common cervical cancer among women in the developing countries, with more than 85 per cent of global cervical cancer-associated deaths occurring in these countries. Molecular epidemiological analysis has shown that HPV infection is the major cause of cervical cancer. More than two hundred HPV genotypes have been described and characterised based on nucleotide sequence relating to the L1 gene, which codes for the major HPV capsid protein. Based on their oncogenic potential via association with cervical cancer and precancerous lesions, HPV have been grouped into two groups: high-risk (HR) genotype, which causes cervical neoplasia, and low-risk (LR) genotype, which causes mild dysplasia. The prevalence of HPV infections in women within the general populace varies considerably within countries and regions, and within regions, ranging from 1.6–41.9 per cent (e.g., a study found that the overall HPV prevalence among Arab women in Qatar was 6.1 per cent).

The human T-cell lymphotropic virus-1 (HTLV-1) is the only retrovirus known to directly cause cancer. HTLV-1 is endemic in south Japan, Central Africa, north-eastern South America, the Caribbean, and Southeastern United States. It is also found among IV drug users in the US and Europe and foci in Middle East and Melanesia. The prevalence of HTLV-1 increases gradually with age, especially among women in all the highly endemic areas. It is transmitted by mother to child, sexual transmission, and transmission through contaminated blood products. Although a lot of data is lacking in large areas, it is estimated that about 5–10 million of people are infected with the virus. HTLV-1 is the major cause of ATL, HTLV-1'-associated myelopathy/tropical spastic paraparesis (HAM/TSP), and uveitis. Due to lack of data, it is important to study HTLV-1 and disease outcomes, such as urinary tract disorders, increased susceptibility to infection, and so on. HTLV-1 has subtypes including subtype A, which includes the prototype sequence from Japan and found mostly in endemic areas worldwide; subtype B, D, and F found in Central Africa; and subtype C found in Melanesia.

KSHV is the cause of KS. The prevalence of KSHV infection varies, from about 1–3 per cent of blood donors in North America to more than 70 per cent in Africa, where it is endemic. The prevalence of KSHV infection approximately mirrors the prevalence of KS. A relatively high seroprevalence of KSHV has been described among IV drug users and women with multiple sex partners, although the incidence of KS among these groups is negligible. KSHV seroprevalence has also been reported to be high among family members of KSHV-seropositive persons. In areas where the virus is endemic, the highest degree of concordant seropositivity is found between mother and child or sibling pairs, and seropositive. Based on these data, the suggestion is vertical or parenteral transmission of KSHV is rare and inefficient, but the high prevalence of the virus among children in most endemic regions also argues against sexual contact as the predominant mode of transmission.

MCV is the cause of approximately 80 per cent of MCC, a rare and highly aggressive skin cancer. It was discovered in 2008 by researchers at the University of Pittsburgh. It also referred to as cutaneous apudoma, primary neuroendocrine, and trabecular carcinoma of the skin. MCC is common among older white men and largely absent in populations younger than forty years. The estimated annual incidence of MCC in the US is about 470 cases per year, but of late, cases of MCC have arisen in organ transplant patients, with reduced immunity and lymphoma. In people with HIV, the relative risk for the tumour is 13.4, when compared to the general population. The tumour is most often located in the sun-exposed skin of the head, neck, and the extremities.

1.3   Do Viruses Cause Cancer? Causality Issues

In medical investigations, the classic standard for measuring causality is based on Koch's postulates. In summary, the postulate states that for a pathogen to be regarded as the cause of a disease, then 1. It should be found in all cases of the disease but not in healthy individuals unless there can be asymptomatic carriers, 2. The pathogen must be isolated from the disease and propagated in culture, 3. Pathogen from the culture should cause the same disease when reinoculated, and 4. The pathogen must be re-isolated from the inoculated host with the disease and be identical to the original agent. These postulates are difficult to apply to human viruses and cancer. First, it is known that in most cases, there is a long period of latency between primary viral infection and the manifestation of cancer. In HTLV-1 infection, for example, the latency period between infection and the onset of acute T-cell leukaemia is to the order of a decade, and in addition, only a fraction of those infected will go on and develop ATL. Also, some viruses establish subclinical infection, so it's difficult to ascertain the time of infection.

Another issue is, oncoviral infection is widespread, but it's rarely associated with cancer. For example, seroepidemiological studies show that 63 to 75 per cent of the population in the US is infected with MCV, but the incidence of MCC is 0.17 to 0.34/100,000. The outcome of viral infection depends on several factors, such as the host's immune status; for example, the immunosuppressive ability of HIV-1 is a major predisposing factor to KSHV. In addition, other viruses require cofactors to initiate the cancer process; for example, in HPV, the cofactors required for cervical cancer include hormonal contraceptive, coinfection with other agents such as chlamydia, smoking, nutrition, and so on. Some oncoviruses such as HSV irreversibly integrate into the host genome during pathogenesis, thereby making it difficult for infectious progeny to be cultured. Most viruses lack an animal model, and some, such as MCV, even lack a cell culture system. Furthermore, other cancer-associated viruses utilise different mechanisms during the process of carcinogenesis; for example, in HPV-associated cervical cancer, the virus promotes chromosomal instability, thereby contributing to cellular genetic changes. The role of some of the viruses in cancer pathogenesis is discussed in later chapters.

So how do we solve this problem? A number of approaches have been suggested based on criteria that define environmental causes, consistency, specificity, temporality, and plausibility. In summary, these guidelines are proposed for a given virus to be regarded as the cause of a human cancer:

1. The geographical distribution of viral infection should match that of cancer after adjustment for other cofactors.
2. Viral markers such as antiviral antibody should be higher in cases of cancer than in controls.
3. Viral markers being present should precede the tumour and have an incidence that matches the incidence of the tumour.
4. Prevention of viral infection though intervention such as vaccination should result in decrease in incidence of the tumour.

5. The virus should exhibit transforming properties with human cells in culture.
6. The virus should induce tumours in animals, and this should be preventable when viral neutralisation techniques are applied.

In general, these factors can be complex, depending on the virus. However, in order to accept these criteria, more research on the virology, epidemiology, and molecular biology of these viruses is needed.

1.4 Viral Carcinogenesis

Molecular biology has played a significant role in the discovery of many mechanisms utilised by oncoviruses to initiate the carcinogenesis processes by altering the function of cellular targets, which plays a significant role in the development of cancer. Carcinogenesis is a multistep process involving pre-initiation, initiation, promotion, and metastasis. With oncoviruses, three mechanisms have been implicated in the carcinogenesis process. These are direct, indirect as a result of chronic inflammation, and indirect through immunosuppression. The direct carcinogens include EBV, HPV, HTLV-1, and KSHV; the indirect carcinogens through chronic inflammation include HBV and HCV; the indirect carcinogens through immunosuppression include HIV-1.

Direct viral carcinogens possess the following characteristics: 1. The entire or partial genome of the virus is usually detected in each cancer cell. 2. The virus expresses a number of oncogenes that interact with cellular proteins to disrupt the checkpoints of cell cycles, inhibit apoptosis and DNA damage response, cause genomic instability, and induce cell immortalisation, transformation, and migration. For example, HCV causes HCC through chronic inflammation, which leads to the production of chemokines, cytokines, and prostaglandins that are secreted by infected cells or inflammatory cells.

Chronic inflammation also results in the production of reactive oxidative species, with direct mutagenic effect to deregulate the immune system, thereby promoting angiogenesis, an essential factor for neovascularisation and survival of tumours. HIV-1-infected individuals are at risk of developing cancers caused by another infectious agent through immunosuppression, leading to increased replication of oncoviruses such as EBV. Although antiretroviral agents have reduced the risk of these cancers, the rate of infection still remains high around the globe.

Only a small portion of those infected with oncoviruses develop cancer. This means there are cofactors which play some roles in the carcinogenesis process of oncoviruses. As stated earlier, carcinogenesis is a multistep process, which means there might be multiple risk factors; for example, viral factors, host factors, and environmental factors. The host and viral factors will be discussed later, but the environmental factors include nutrients, immunosuppression drugs, and co-infection with other pathogens. Other factors might be involved but are yet to be identified. Alteration of certain genes (through mutation) involved in cellular functions leads to malignant transformation; the best documented is the tumour suppressor protein p53, which will be dealt with in detail in chapter 6. Several viral oncogenic proteins and factors have been described. This section will highlight a few of them. Other proteins associated with such processes will be discussed later.

1.4.1 Hypoxia-Inducible Factor 1 (HIF-1)

HIF proteins are a major component of the innate hypoxic stress response in non-cancerous cells, acting as multitude of genes needed for adaptation under low oxygen. So far, three HIF isoforms has been described, viz. HIF-1, HIF-2, and HIF-3. Available data shows that activation of HIV-1 transcription factor is a pathway mostly affected by human oncoviruses. Bersten et al. described the component of HIF-1. Briefly, it is a heterodimer consisting of α and β subunits. The dimer is a member of helix loop helix-PER-ARNT-SIM (bHCH-PAS), a family of transcription factors associated with the development of cancer. In normal, non-hypoxic cells, HIF-1α is synthesised continually and degraded, while HIF-1β is also continuously expressed to levels that remain constant with the nucleus. HIF-1 activity is therefore dependent on the regulation of HIF-1α. HIF-1α mRNA is expressed, and its levels are similar for most cells studied between hypoxic and normoxic condition. However, in some cell types, such as HCC Hep3B cells, there is an increase in HIF-1α transcription under hypoxic conditions. Other *in vivo* studies showed that there is the potential of environmental hypoxia inducing HIF-1α transcription. Therefore, most studies support the suggestion that oncoviruses enhance HIF-1α levels through the modulation of its transcription, translation, or stabilisation. However, evidence is lacking on whether HIF-1α target gene is necessary for malignant transformation. Also, a complete host-virus interaction, effects of viral genomic variation, and the potential therapeutic benefits of utilising HIF-1α in viral carcinogenesis require more investigations.

1.4.2   Proto-Oncogene

An oncogene is a gene that codes for a protein that can potentially transfer a normal cell into a malignant cell. Most normal cells undergo apoptosis (a process of cell death) when critical functions are altered. An activated oncogene causes a cell designated for apoptosis to survive and proliferate instead. Oncogenes are normally influenced by external factors such as viruses or environmental conditions. A proto-oncogene is a normal gene that can become an oncogene due to mutation, which leads to increase in protein expression, hyperactivity, and loss of regulation. Proto-oncogenes are often involved in signal transduction and have mitogenic effects. Upon activation, a proto-oncogene or its product becomes a tumour-inducing agent. A number of proto-oncogenes have been associated with malignancy. These include RAS, MYC, WNT, ERK, E6, E7, and TRK. MYC gene is implicated in Burkitt's lymphoma. Some of the proto-oncogenes will be dealt with in chapter 5.

A proto-oncogene can become activated by

- point mutation,
- amplification,
- translocation to a transcriptionally active site, or
- chimeric gene creation due to chromosomal rearrangements.

1.4.3   Tumour Suppressor Genes

A tumour suppressor gene protects a cell from the first step of the cancer process. Mutation in this gene can cause loss or reduction of its function. The cell then progresses to cancer, usually in combination with other factors. The tumour suppressor gene has a number of functions, including the following:

- repression of genes that is essential for cell cycle
- coupling the cell cycle to DNA damage
- initiation of apoptosis
- metastasis suppression

Examples of tumour suppressor genes include Retinoblastoma protein (pRB), found in human retinoblastoma, and p53, which is encoded by TR53 gene. P53 as a tumour suppressor gene will be discussed in detail in chapter 6.

The main weapon that oncoviruses deploy for cancer pathogenesis is persistence. In order to achieve that, they utilise their ability to evade the immune system. The effect of host immunity in the pathogenesis of oncoviruses will be dealt with in chapter 2.

Some hormones are known to be important in the development of cancer by promoting cell proliferation. Insulin-like growth factors and their binding protein play a key role in cancer cell proliferation, differentiation, and apoptosis. But do hormones play a role in the pathogenesis of viral-associated cancer? A number of studies have been undertaken to analyse the effect of hormones in the development of viral-associated cancer. Lindström and Hellberg investigated the expression of leucin-rich repeats and immunoglobin-like domains 3 (LRIG3) in invasive cancer and cervical intraepithelial neoplasm (CIN) for possible correlation with other tumour markers; in their study involving 129 patients with invasive squamous cell carcinoma and 170 biopsies showing high and low grade CIN, or normal epithelium, found that in CIN, there was high expression of the tumour suppressors retinoblastoma, p53, and p16, and E-cardherin or low expression of CK10, correlated to LRIG3 expression. In addition, progestogenic contraceptive use correlated with high expression of LRIG. High LRIG expression correlated significantly with the presence of high-risk HPV infection in patients with normal epithelium or CIN. A study by Lauttia et al. to analyse the effect of prokineticins in MCC with Merkel cell polyomavirus infection concluded that prokineticins are associated with Merkel cell polyomavirus infection and participate in regulation of the immune response in MCC, and this may influence the outcome of MCC patients. The prokineticins family are chemokine-like proteins that are highly conserved across species. Data showed that they have influence in great diversity of biological functions and participate in the coordination of complex physiological activities such as feeding, drinking, regulation of circadian rhythm, and hyperalgesis, and they are suspected of being involved in angiogenesis, inflammation, and cancer. However, a study found that relationship between HCC and metabolic factors other than diabetes is inconclusive, although another study found that triglyceride levels were inversely associated with subsequent HBV-associated HCC. Hormones like estradiol 2 (E2) are confirmed cofactors for HPV-associated cervical cancer. Most of the studies analysing the effect of hormones on viral-associated cancer were inconclusive; therefore, more studies are needed.

1.4.4   MiRNA and Oncogenesis

Micro (mi) RNAs are small, noncoding, highly stable 22-base pair nucleic acids that were first discovered in the nematode *Caenorhaditis elegans*. They are mobile and functional genetic elements. miRNA genes play important roles in developmental timing, morphologic changes, cell proliferation and death, hematopoiesis, nervous system control, pancreatic insulin secretions, adipogenesis, oncogenesis, and viral diseases. A typical miRNA gene is made of 5'-terminal monophosphate and 2',3'-diol at their 3'-terminus (although some modifications have been reported) Unlike mRNA, which serves as a template for DNA translation to protein, miRNA are unique in the sense that they directly regulate this translation. Several hundreds of these oligomers have been discovered.

Viruses produce their own set of miRNAs which have been implicated in virus-associated carcinogenesis. The first to be identified as regulatory viral miRNA was miR-S1 in simian virus 40 (SV40), which promotes the recognition and destruction of infected cells by cytotoxic T cells, while the first reported trans-regulatory viral miRNA was miR-LAT, which targets TGF-β and SMAD3, thereby promoting cellular proliferation and preventing apoptosis in HSV-1 infection. A number of studies have tried elucidating the role of miRNA in viral-associated cancers. In EBV infection, miRNA was first described by Pfeffer et al.; since then, forty-four mature miRNAs from precursors miRNA have been discovered. Studies have found that they are encoded in two regions: BART and BHRF1. They show variable expression in cell lines and tumours during viral latency and lytic growth. Furthermore, it has been found that EBV miRNA can change host miRNA expression. EBV miRNA have been implicated in regulating host cell proapoptotic proteins BBC3/PUMA and BCL2LII/BIM, viral transcripts and transport, as well as immunomodulatory targets. The BHRF1 miRNA has been implicated in the inhibition of apoptosis during initial infection and in cell cycles. BHRF1 miRNA are expressed in all forms of EBV latency and also target mRNA transcript involved in host cell apoptosis and cell cycling. Due to these factors, it has been suggested that miRNA-regulated post-transcriptional regulation of host mRNA may be vital for virus-mediated host cell malignant transformation. Some interesting miRNA-mRNA interactions have been described. These include miR-BARTs-targeting BALF5 (a viral polymerase) and several BART clusters such as miRNAs downregulating the oncogenic late membrane protein (LMP) 1 and miR-BART-22, which targets LMP2. BART miRNAs are associated with downregulation of proapoptotic and tumour suppressor cellular targets in NPC such as PUMA, WIF1, and APC. In HIV-associated cancers, it has been shown that HIV-1 miRNA is found in the U3 region of the 3'-LTR, and it downregulates cellular apoptosis antagonising transcription factor (AATF) gene expression. The AATF interacts with POL II and the tumour suppressor pRB. It has also been reported that AATF is associated with endogenous antagonist of prostrate apopotosis-4 (Par-4), which is reported to be associated with the suppression of Bcl2 gene transcription. HIV-1 miR-HI is likely to activate E2F activity and inhibit apoptosis. This might be the co-factor in the carcinogenesis process of HIV-1 infection. In HTLV infection, Pichler et al. and Yeung et al. reported that miRNA is either directly activated by Tax (a transactivator oncoprotein) or associated with HTLV-1-induced cell transformation. In their study, Pichler et al. selected a limited number of miRNAs with links to cancer and overexpressed in regulatory T lymphocytes. RT-PCR quantification analysis identified upregulated and repressed miRNAs in cell lines derived from ATL patients, HAM/TSP patients, and HTLV-1- or Tax-transformed cells. The upregulated miRNAs were miR-21, miR-24, miR-146a, and miR-158, while the repressed was miR-223. Expression of one of the miRNAs (miR-146a) was directly activated by Tax through the proximal NF-kB site of the MIRNA146A gene promoter.

Yeung et al., on the other hand, profiled 327 human miRNAs in seven HTLV-1 transformed cell lines and four from PBMC samples from acute ATL patients. Among fifteen miRNAs whose expression was consistently modified compared to paired controls, only three (miR-93, miR-130b, and miR-18a) were induced upon activation of normal PBMC with phorbol myristate acetate. By using RT-PCR analysis, the differential expression of miR-93 and miR-130b was confirmed by the authors. Furthermore, luciferase reporter assays and computational analysis showed that p53-induced tumour suppressor protein (TP53INP1) was a target shared by both miR-93 and miR-130b. Utilising antagomirs for miR-93 and miR-130b restored the expression of TP53INP1 and increased the apoptosis of HTLV-1-transformed MT4 cells. siRNA knock-down of TP53NPI1 saved MT4 from cell death induced by miR-130b antagomirs. Increased expression of miR-130b was partly associated with transcriptional activation by Tax.

A number of miRNAs have been reported in ATL cells. This means HTLV-1 may either subvert or include cellular miRNAs for persistence and transformation. However, because of lack of conclusive data, more studies are needed to elucidate the role of miRNAs in HTLV-associated oncogenesis and identify more miRNA for proper diagnosis and management of HTLV- associated diseases. miRNAs are essential components of the viral oncogenetic process. Understanding their roles will aid in formulating therapeutic drugs and biomarkers of oncoviral diseases.

1.4.4 Epigenetic

Epigenetic was defined by Robin Holiday as heritable changes in gene expression that do not result from any alteration in the DNA sequence. It includes the methylation status of DNA and the post-translational modification of histones. The epigenetic process plays an important function in tumorigenesis in mammals. The best known marker of epigenetic is DNA methylation, which together with specific histone modification and specific miRNA are believed to be a defining molecular landscape that is altered in cancer. DNA methylation occurs mainly on the cytosine, that is, before guanine, resulting in the formation of 5-methlycytosine. These dinucleotide sites are referred to as CpGs. The CpGs are found in the promotor regions of about half of all genes. They are called the CpG Island because they are distributed asymmetrically into the CpG-poor region and other areas. These CpG Islands are usually unmethylated in the normal cells, while sporadic CpG sites in the rest of the genome are normally methylated. DNA methylation is a normal process in cells of the mammals that allows normal expression patterns to be maintained. It is involved in genomic imprinting, X-chromosome inactivation in females, and silencing parasite as well as foreign elements, among others. But methylation of CpG Islands in the promoter region is associated with gene silencing and aberrant DNA methylation, as reported in most cancers leading to the silencing of some tumour suppressor genes.

As explained earlier, oncoviral genomes disrupt the host genome by inserting mutations and chromosome rearrangement, predisposing the infected cells to cancer. It is also known that viral genes are associated with aberrant methylation profiles in host-specific genes in human cancers. A number of studies have attempted to elucidate the epigenetic changes in the viruses. In EBV, epigenetic regulation of viral genes is essential event in the life cycle of the virus. The expression of latent viral oncogenes, RNA, and miRNA is under epigenetic control by DNA methylation and histone modification, resulting in complete silencing of the EBV gene.

Methylation of EBV genome protects the host cell through subduing the transforming latency gene. However, the virus uses DNA methylation to maximise its persistence strategies and hide itself from immune detection, inhibiting the expression of viral latency proteins which are recognised by cytotoxic T cells.

HPV infection has also been associated with viral and host epigenetic modification involving DNA methylation and histone modification, which contribute to pathogenesis and tumorigenesis. Viral DNA methylation is more associated with carcinoma than asymptomatic infections or dysplasia. For example, HPV 16 and 18, the LCR and E6 sequences, were reported to be unmethylated regardless of the stage of neoplastic progression, while the L1 region was densely methylated. Methylation studies have shown that in HPV16, the LCR was methylated in some primary cervical carcinoma, especially at E2-binding sites (E2BS). E2BS have been shown to inhibit the binding of E2; methylation is related to activation of E6 and E7 viral proteins.

Similarly, epigenetic changes have been described in HCV infection. It has been reported that this epigenetic alteration is produced by viral proteins. HBx protein and various HBs envelope proteins are responsible for the alteration of major signaling pathways. HBx is the key factor in initiating epigenetic alteration induced by the virus. HBx interacts with DNMT to initiate the epigenetic alteration. HBV replication is also associated with epigenetic markers such as the acetylation of H3 and H4. Epigenetic alterations have been described in almost every known oncovirus. More studies are needed for us to understand the epigenetic process better because it will give us the option of developing better and more effective chemotherapy.

1.5 Molecular Tools for Research in Oncoviral Infections

Advances in molecular techniques have enhanced our current understanding of the carcinogenetic process involved in viral-associated cancer development. Here is a summary of the current molecular tools utilised in viral-associated cancer research.

1.5.1   Polymerase Chain Reaction Assay

Most molecular-based studies begin with the extraction of DNA from a particular organism, followed by the amplification (i.e., generation of many copies) of particular segments of DNA using the polymerase chain reaction (PCR). PCR is used because only minute quantities of DNA are needed (e.g., nanogram amounts). Figure 1 describes the various stages of a typical PCR technique.

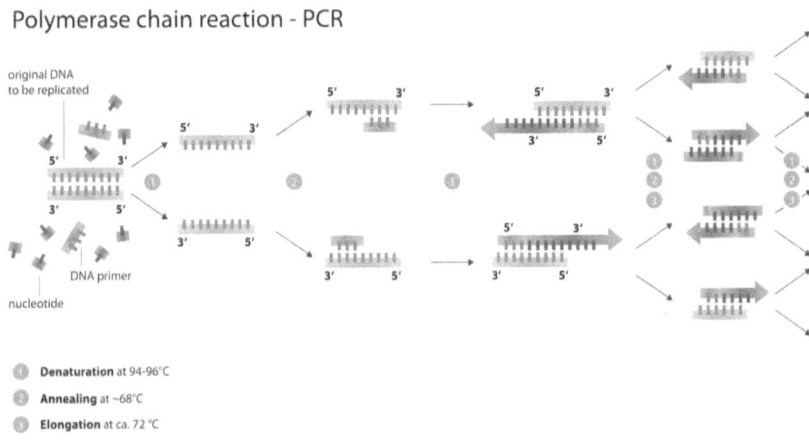

Figure 1.1: The PCR method begins with total genomic DNA extracted from an organism. The DNA is combined with site-specific primers, taq polymerase, and other reagents (e.g., MgCl2, buffer, dNTPs) and subjected to repeated cycles, each of which consists of a denaturation phase, annealing phase, and elongation phase. Denaturation separates double-stranded DNA, allowing primers to anneal to specific sites, followed by incorporation of deoxynucleotide triphosphates (dNTPs; A, C, G, T), thereby extending the target site in the 5'-3' direction (on both separated strands). The first cycle is completed when one round of denaturation, annealing, and elongation is finished, resulting in two new copies of the target site. Subsequent cycles (typically, thirty to thirty-five) repeat the three-phase process, resulting in many million-fold copies of amplified DNA.

Recent advances in biotechnology and molecular biology ha^ve led to the development of a variety of PCR assay such as real-time PCR, PCR-ELISA, quantitative-PCR, and multiplex-PCR, which can help in identification and also characterisation of various pathogens, such as oncoviruses.

1.5.2   Markers and Methods

There are many types of genome markers used in molecular virology; these include microsatellite, minisatellite, restriction fragment length polymorphism (RFLP), and genome sequence data. Markers generated by these methods can also be visualised in different ways. Traditionally, microsatellite and RPLFs are visualised as discrete bands, revealed by agarose electrophoresis. The genome analysis can be visualised using polyacrilamide gels and autoradiography. Due to advancements in molecular techniques, these markers can also be visualised using chemiflourescence and genetic analysers, which detect fluorescent emission of a labeled primer or fluorescently labeled nucleotide of a genome sequence. These markers and visualisation methods are by no means comprehensive, and the techniques to be utilised depend greatly on the question involved. But understanding different types of information provided by different markers can assist in deciding which is best for a particular study. A marker is a biological indicator that signals a change in the physiological state, stress, or injury due to disease or the environment.

There are three types of markers. Anonymous markers include those generated by a method called amplification fragment length polymorphism (AFLP). This technique uses restriction enzymes combined with PCR to generate many unique fragments that can be used to distinguish organisms based on their genomes. The technique involves three steps: 1. Restriction of the DNA and ligation of oligonucleotide adapter. 2. Selective amplification of set of restriction fragments. 3. Gel analysis of the amplified fragments. With this method, sets of restriction fragments may be visualised by PCR without knowledge of the nucleotide sequence, which allows for specific co-amplification of a high number of restriction fragments. A disadvantage of AFLPs is they are limited in the type of information they provide. Another similar method is called random amplified polymorphism DNA

(RAPD), which just like AFLP generates dominant markers that can be visualised using agarose gel electrophoresis. RAPD uses PCR primers set to randomly amplify DNA fragments scattered throughout the genome.

Another class of markers is sequence-tagged site (STS) markers, which is a short (200–500bp) DNA sequence that has a single occurrence in the genome, whose location and base sequence are known. STS can easily be detected with PCR using specific primers designed by the investigator. One type of STS is simple sequence repeats (SSRs) or variable number tandem repeats (VNTRs). Unlike AFLP, a prior knowledge of a specific region containing tandemly repeating nucleotide motifs is required. It focuses on microsatellite regions of the genome. Most SSRs consist of two or three nucleotides. The number of SSRs within a given microsatellite region of the genome often varies among individuals.

The third class of markers used by molecular virologists is Sanger sequencing, in which the markers are derived from direct DNA sequencing of the targeted region within the genome. Just like STS, DNA sequencing requires precise knowledge of the specific genes or gene region that is of interest to the investigator. This technique is ideal for studying evolutionary trends of viruses. DNA sequencing also enables the development of a marker type, single nucleotide polymorphisms (SNPs). When multiple sequences of a particular region are generated for multiple members within a species, a single base difference among individuals is often detected. Another approach to targeting individual genes is whole genome sequencing. A new method that rapidly generates sequences that can be analysed and compiled into a whole genome is next generation sequencing (NGS), which is now becoming an important tool for molecular virologists interested in investigating the entire genome of a particular virus. All the molecular methods have their strengths and weakness; therefore, choosing a particular method depends on the research aim.

1.5.3 Which Experimental Method Suits My Research?

A key question to ask when thinking of utilising molecular techniques in oncoviral research is, which is best suited for my research? The answer to this question will be determined by a number of factors, all of which must be evaluated both independently and collectively in order to arrive at a cohesive plan before launching a successful molecular study in oncoviral research.

1. Are you interested in characterising a specific protein or gene involved in the pathogenesis of a particular oncovirus?
2. Do you want to investigate immune response and immunity in viral-associated cancer?
3. Do you want to develop and evaluate vaccines, host-viral interactions, and nonhuman primate models?
4. Do you want to develop novel antiviral drugs that are effective against oncoviruses?
5. Do you want to evaluate a rapid diagnostic test?
6. Are you interested in constructing evolutionary trends of a particular oncovirus?

There are multiple methods to use in assessing your question, but thinking carefully about what you want to achieve will determine how you answer your question and address it in the field of molecular oncovirology.

REFERENCES

Agelli M, Clegg LX (2003), Epidemiology of primary Merkel carcinoma in the United States. J Am Acad Dermatol 49:832–841.

Akkina R (2013), Human immune response and potential for vaccine assessment in humanized mice. Curr Opin Immunol 25:403–409.

Ambos V (2004), The function of animal miRNAs. Nature 451: 350–355.

Angeloni A, Heston L, et al. (1998), High prevalence of antibodies to human herpesvirus 8 in relatives of patients with classic Kaposi's sarcoma from Sardina. J Infect Dis 1715–1718.

Baltimore DL (1970), RNA-dependent DNA polymerase in virion of RNA tumor viruses. Nature 226: 1209–1211.

Bansal D, Elmi AA, et al. (2014), Molecular epidemiology and genotype distribution of Human Papillomavirus (HPV) among Arab women in the state of Qatar. J of Transl Med 12:300.

Barbera AJ, Ballestas ME, Kaye KM (2004), The Kaposi's sarcoma-associated herpesvirus latency-associated nuclear antigen 1 N terminus is essential for chromosome association, DNA replication, and episome persistence. J Virol 78(1):294–301.

Barre-Sinoussi F, et al. (1983), Isolation of a T-lymphotropic retrovirus from a patient at risk for acquired immune deficiency syndrome (AIDS). Science 220: 808–871.

Barth S, et al (2008), Epstein-Barr virus-encoded microRNAs miR-BART2 downregulates the virual DNA polymerase BALF5. Nucleic Acid Res 36: 666–675.

Beasley RP, Hwang LY, et al (1981), Hepatocellular carcinoma and hepatitis B virus: A prospective study of 22,707 men in Taiwan. Lancet 2:1129–1133.

Becsei-Kilborn E (2010), Scientific discovery and scientific reputation: The reception of Peyton Rous' discovery of the chicken sarcoma virus. J Hist Biol 43:111–157.

Bersten DC, Sullivan AE, et al (2013), bHLH-PAS proteins in cancer. Nat Rev Cancer 13: 827–841.

Bility MT, Li F, et al (2013), Liver immune-pathogenesis and therapy for human liver tropic virus infection in humanized mouse models. J Gastroenterol Hepatol 28 (Suppl1): 120–124.

Bittner JJ (1936), Some possible effects of nursing on the mammary gland tumor incidence in mice. Science 84:162.

Blumerg BS, Alter HJ, Visni US (1965), A new antigen in leukemia sera. JAMA 191:541–546.

Boccardo E, Villa LL (2007), Viral antigen of human cancers, Curr Med Chem 14:2526–2539.

Bouvard V, Bann R, et al (2009), A review of human carcinogens, Part B: Biological agents. The Lancet Oncology 10:321–322.

Bouzer AB, Willems L (2008), How HTLV-1 may subvert miRNAs for persistence and transformation. Retrovirology 5: 101.

Boyle P, Levin B (2008), International Agency for Research on Cancer World Health Report. Lyon, Geneva: International Agency for Research on Cancer. Distributed by WHO Press.

Burke W, Daly W, Garber J (1997), Recommendation for follow-up care of individuals with inherited predisposition to II. BRACA 1 and BRACA 2. J of Am Med Assoc 277:997–1003.

Cai X, et al (2006), Epstein-Barr virus microRNAs are evolutionarily conserved and differentially and expressed. PLos Pathog 2: e23.

Cannon MJ, Dollard SC, et al (2001), Blood-borne and sexual transmission of human herpesvirus 8 in women with or at risk for human immunodeficiency virus infection. NEJM 344:637–643.

Cheema SK, et al (2003), Par-4 transcriptionally regulates Bcl-2 through a WTI-binding site on the bcl-promoter. J BIol Chem 278: 19995–20005.

Chen C-J, Hsu W-L, et al (2014), Epidemiology of virus infection and human cancer. Recent Results Cancer Research DOI: 10.1007/978-3-642-38965-8_2.

Chen C-J, Yang HI, et al (2008), Metabolic factors and risk of hepatocellular carcinoma by chronic hepatitis B/C infection: A follow-up study in Taiwan. Gastroenterology 135:111–121.

Chen CZ, et al (2004), MicroRNAs modulate hematopoietic lineage differentiation. Science 303:83–86.

Chen J (2008), Is the Src the key to understanding metastasis and developing new treatment for colon cancer? Nat Clin Pract Gasteroenterol Hepatol 5:306–307.

Chen XM (2004), MicroRNAs as a translational repressor of APETALA 2 in Arabidopsis flower development. Science 303: 2022–2025.

Chiang C-H, Huang K-C (2014), Association between metabolic factors and chronic hepatitis B virus. World J Gastrolenterol 20:7213–7216.

Choy EY, et al (2008), An Epstein-Barr virus-encoded microRNA targets PUMA to promote host cell survival, J Exp Med; 205: 2551–2560.

Chuang JC, Jones PA (2007), Epigenetics and microRNAs. Pediatr Res 61: 24R–29R.

Clifford GM, Polesel J, et al (2005), Cancer risk in Swiss HIV cohort study: Association with immunodeficiency, smoking, and highly active antiretroviral therapy. J Natl Cancer Inst 97:425–432.

Cuninghame S, Jackson R, Zehbe I (2014), Hypoxia-inducible factor 1 and its role in viral carcinogenesis. Virology 456-457:370–383.

Datta S (2008), An overview of molecular epidemiology of hepatitis B virus (HBV) in India. Virology J 5:156.

De Martel C, Ferley J, et al (2008), Global burden of cancers attributable to infections in 2008: A review and synthetic analysis. The Lancet Oncology 13:607–615.

Dolken L, et al (2010), Systematic analysis of viral and cellular microRNAs targets in cells latently infected with human γ-herpesviruses. Cell Host Membrane 7:324–334.

de Martel C, Francesch S (2009), Infections and cancer: Established association and new hypothesis. Crit Rev Oncol Hematol 70:183–194.

Duncavage EJ, Zehnbauer BA, Feifer JD (2009), Prevalence of Merkel cell polyomavirus in Merkel cell carcinoma. Modern Pathology 22:516–521.

Eddy BE, Borman GS, et al (1962), Identification of the oncogenic substance in rhesus monkey kidney cells as Simian virus 40. Virology 17:65–75.

Ellerman V, Bang O (1908), Experimental leukemia bei huhaern. Centralbl.f. Bakteriol; 46:595–609.

Engels EA, Frisch M, et al (2002), Merkel cell carcinoma and HIV infection. Lancet. 359:497–498.

Epstein MA, Achong BG, Barr YM (1964), Virus particle in controlled lymphoblast from Burkitt's lymphoma. Lancet 15: 702–703.

Esteller M (2008), Epigenetics in cancer. N Engl J Med 358: 1148–1159.

Evans AS, Mueller NE (1990), Viruses and cancer: causal association. Ann Epidemiol 1:71–92.

Feinberg AP, Cui H, Ohlsson R (2002), DNA methylation and genomic imprinting: Insights from cancer into epigenetic mechanisms. Semin Cancer Biol 12: 389–398.

Feng H, Taylor JL, et al (2007), Human transcriptome subtraction by using short sequences tags to search for tumor virus in conjunctival carcinoma. J Virol 81:11352–11340

Feng H et al (2008), Clonal integration of a polyomavirus in human Merkel cell carcinoma. Science 319:1096–1100.

Fernandez AF, Rosales C, et al. (2009), The dynamic DNA methylomes of double-stranded DNA viruses associated with human cancer. Genome Res 19: 438–451.

Fraga MF, Esteller M (2005), Towards the human cancer epigenome: A first draft of histone modifications. Cell Cycle 4: 1377–1381.

Fuji YR (2008), Formulation of new algorithm for miRNAs. Open Virol J 2: 37–43.

Gariglio P, Gutierrez J, et al (2009), The role of retinoid deficiency and estrogens as cofactor of cervical cancer. Arch Med Res 40:449–465.

Gatti G, et al (2008), Epstein-Barr virus latent membrane protein I trans-activates miR-155 transcription through the NF-kB pathway. Nucleic Acid 36:6608–6619.

Gessain A, Cassar O (2013), Epidemiological aspects and world distribution of HTLV-1 infection. Front Microbiol 3:388.

Gross L (1951), "Spontaneous" leukemia developing in infancy, with AK-leukemic extracts, or AK-embryos. Proc Soc Exp Biol Med 76:21–32.

Gupta A, et al (2006), Anti-apoptotic function of a microRNAs-encoded by the HSV-1 latency-associated transcript. Nature 442: 82–85.

Gurtsevitch VE (2008), Human oncogenic viruses: Hepatitis B and hepatitis C viruses and their role in hepatocarcinogenesis. Biochemistry (Mosc) 73: 504–513.

Henle W, Diehl V, et al (1967), Herpes-types virus and chromosome marker in normal leukocytes after growth with irradiated Burkett cells. Science 157:1064–1065.

Herman JG, Baylin SB (2003), Gene silencing in cancer in association with promoter hypermethylation. NEJM 349: 2042–2054.

Hill AB (1965), The environment and disease: Association or causation? Proc R Soc Med 58: 295–300.

Hodgson NC (2005), Merkel cell carcinoma: Changing incidence trends. J Surg Oncol 89:1–4.

Holliday R (1987), The inheritance of epigenetic defects. Science 238: 163–170.

Huang J, et al (2007), Cellular microRNAs contribute to HIV-1 latency in resting primary CD4+ T lymphocytes. Nat Med 13: 1241–1247.

Hublarova P, Hrstka R, et al (2009), Prediction of human papillomavirus 16 e6 gene expression and cervical intraepithelial neoplasia progression by methylation status. Int J Gynecol Cancer 19: 321–325.

Hu J, Garber AC, Renne R (2002), The latency-associated nuclear antigen of Kaposi's sarcoma-associated herpesvirus supports latent DNA replication in dividing cells. J Virol 76:11677–11687.

IARC (1995), *Human Papillomavirus*, Volume 64, Lyon, France: IARC.

Jemal A, Bray F, et al (2011), Global cancer statistics. CA Cancer J Clin 61:69–90.

Jones PA, Baylin SB (2002), The fundamental role of epigenetic events in cancer. Nat Rev Genet 3: 415–428.

Jones PA, Takai D (2001), The role of DNA methylation in mammalian epigenetics. Science 293: 1068–1070.

Jung JK, et al (2007), Expression of DNA Methyltransferase 1 is activated by hepatitis B Virus X protein via a regulatory circuit involving the p16INK4a-cyclin D1-CDK 4/6-pRb-E2F1 pathway. Cancer Res 67: 5771–5778.

Kaposi's sarcoma and Pneumocystis pneumonia among homosexual men in New York City and California. MWWR Morb Mortal Wkly Rep 30:305–308.

Kelley-Clarke B, BallestasME, et al (2007), Kaposi's sarcoma herpesvirus C-terminal LANA concentrates at pericentromeric and peri-telomeric regions of a subset of mitotic chromosomes. Virology 357:149–157.

Kelley-Clarke B, et al (2007), Determination of Kaposi's sarcoma-associated herpesvirus C-terminal latency-associated nuclear antigen residues mediating chromosome association and DNA binding. J Virol 81 :4348–4356.

Kozomara A, Griffths-Jones S (2011), miRBase: Integrating microRNAs annotation and deep-sequencing data. Nucleic Acid 39 (Database issue), D152–D157.

Korzeniewski N, Spendy N, et al (2011), Genomic instability and cancer: Lesson learned from human papillomavirus. Cancer Lett 305:113–122.

Krichevsky AM, et al (2003), A microRNAs array reveals extensive regulation of microRNAs during brain development. RNA 9: 1274–1281.

Krithivas A, Fujimuro M, et al (2002), Protein interactions targeting the latency-associated nuclear antigen of Kaposi's sarcoma-associated herpesvirus to cell chromosomes. J Virol 76:11596–11604.

Lai H-C, Lin Y-W, Huang THM, et al (2008), Identification of novel DNA methylation markers in cervical cancer. Int J Cancer 123: 161–167.

Lairmore MD, Haines R, Anupam R (2012), Mechanism of human T-lymphotorphic virus type 1 transmission and disease. Curr Opin Virol 2: 474–481.

Lane DP, Crawford LV (1979), T antigen is bond to a host protein in SV40-transcribed cells. Nature 278:261–263.

Lauttia S, Sinto H, et al (2014), Prokineticins and Merkel cell polyomavirus infection in Merkel cell carcinoma. Br J Cancer 110:1446–1455.

Lindström AK, Helberg D (2014), Immunohistochemical LRIG3 expression in cervical intraepithelial neoplasm and invasive squamous cell cervical cancer: Association with expression of tumor markers, hormones, high-risk HPV infection, smoking and patient outcome. Eur J Histochem 58:2227.

Linn Staedt SD, et al (2010), Virally induced cellular microRNAs miR-155 plays a key role in B-cell immortalization by Epstein-Barr virus. J Virol 84: 11670–11678.

Linzer DI, Levine AJ (1979), Characterization of a 54k Dalton cellular SV40 tumor antigen present in SV40-transcribed cells and uninfected embryonal carcinoma cells. Cell 17:43–52.

Lu AK, et al (2007), Modulation of LMP1 protein express by EBV-encoded microRNAs. PNAS USA 104:16164–16169.

Lu F, et al (2008), Epstein-Barr virus-induced miR-155 attenuates NF-kB signaling and stabilizes latent virus persistence. J Virol 82: 10436–10443.

Lujambio A, Ropero S, et al (2007), Genetic unmasking of an epigenetically silenced microRNA in human cancer cells. Cancer Res 67: 1424–1429.

Lung RW, et al (2009), Modulation of LMP2 expression by a newly identified Epstein-Barr virus-encoded microRNAs miR-BART22. Neoplasia 11: 1174–1184.

Lupberger J, Hildt E (2007), Hepatitis B virus-induced oncogenesis. World J Gastroenterol 13: 74–81.

Marquitz AR, et al (2011), The Epstein-Barr virus BART microRNAs target the pro-apoptotic protein Bim. Virology 412: 392–400.

Monaco L, et al (2003), Genomic structure and transcriptional regulation of che-1: A novel partner of Rb. Gene 321:57–63.

Monner J, Samsor M (2010), Prokineticins in angiogenesis and cancer. Cancer Lett 296:144–149.

Munoz N, Bosch FX, et al (2003), Epidemiologic classification of human papillomavirus types associated with cervical cancer. NEJM 348:518–527.

Munoz N, Castellsague X, et al (2006), HPV in the etiology of human cancer. Vaccine 24 (Suppl 3),S3/1–10.

Niller HH, Wolf H, Minarovits J (2008), Regulation and dysregulation of Epstein-Barr virus latency: Implications for the development of autoimmune diseases. Autoimmunity 41: 298–328.

Pagano JS (1999), Epstein-Barr virus, the first human virus and its role in cancer. Proc Assoc Am Physicians 111:573–580.

Pagano JS, Blaser M, et al (2004), Infectious agent and cancer: Criteria for a causal relation. Semin Cancer Biol 14:453–471.

Park J-J, Kim Y-E, et al (2007), Functional interaction of the human cytomegalovirus IE2 protein with histone deacetylase 2 in infected human fibroblasts. J Gen Virol 88: 3214–3223.

Parker RS, Varmus Bishop JM (1981), Cellular homologue (c-src) of the transforming gene of Rous sarcoma: Isolation, mapping, and transcriptional analysis of c-src and flanked regions. PNAS USA 78:5842–5846.

Paulson EJ, Speck SH (1999), Differential methylation of Epstein-Barr virus latency promoters facilitates viral persistence in healthy seropositive individuals. J Virol 73: 9959–9968.

Payer B, Lee JT (2008), X chromosome dosage compensation: How mammals keep the balance. Ann Rev Genet 42: 733–772.

Penn I (1999), Merkel's cell carcinoma in organ recipients: Report of 41 cases. Transplantation 68:1717–1721.

Persing DH, Prendergast FG (1999), Infection, immunity and cancer. Arch Pathol Lab Med 123:1015–1022.

Pfeffer S, et al (2004), Identification of virus-encoded microRNAs. Science 304: 734–736.

Pfeffer S, Voinnet O (2006), Viruses, microRNAs and cancer. Oncogene 25: 6211–6219.

Phelan JA (2003), Viruses and neoplastic growth. Den Clin North Am 47:533–543.

Pichler K, et al (2008), MicroRNA miR-146a and further oncogenesis-related cellular microRNAs are dysregulated in HTLV-transformed T lymphocytes. Retrovirology 5:100.

Plancoulaine S, Abel L, et al (2000), Human herpesvirus 8 transmission from mother to child and between siblings in an endemic population. Lancet 356:1062–1065.

Poiesz BJ, Ruscetti FW, et al (1980), Detection and isolation of type C retrovirus particles from fresh and cultured lymphocytes of a patient with cutaneous T-cell lymphoma. PNAS USA 77:7415–7419.

Pollicino T, Belloni L, et al (2006), Hepatitis B virus replication is regulated by the acetylation status of hepatitis B virus cccDNA-bound H3 and H4 histones. Gastroenterology 130: 823–837.

Prince AM, Fuji H, Gershan RK (1964), Immunochemical studies on the etiology of anicteric hepatitis in Korea. Am J Hyg 79: 365–381.

Rahadiani N, et al (2008), Late membrane protein-1 Epstein-Barr virus induces the expression of B-cell integration cluster, a precursor form of microRNAs-155, in B lymphoma cell lines. Biochem Biophys Res comm 377: 579–583.

Reinhart BJ, et al (2000), The 21-nucleotide let-7 RNA regulates developmental timing in *Caenorhabditis elegans*. Nature 403:901–906

Rous PA (1910), A transmissible avian neoplasm (Sarcoma of the common fowl). J. Exp. Med 12:696–705.

Rous PA (1911), A sarcoma of the fowl transmissible by an agent separable from the tumor cells. J Exp Med 13:397–411.

Rubel JR, Milford EL, Abdi R (2002), Cutaneous neoplasms in renal transplant recipients. Eur J Dermatol 12:532–535.

Schafer G, Blumenthal MJ, Katz AA (2015), Interaction of human tumor viruses with host cell surface receptors and cell entry. Viruses 7:2592–2617.

Schiffman M, Castle PE (2003), Human papillomavirus: Epidemiology and public health. Arch Path Lab 127:930–934.

Schiller JT, Lowy DR (2010), Vaccines to prevent infections by oncoviruses. Am Rev Microbiol 64:23–41.

Seiki M, Hatton S, et al (1993), Human adult T-cell leukemia virus: Complete nucleotide sequence of the provirus genome integrated in leukemia cell DNA. PNAS USA 80: 3618–3622.

Seto E, et al (2010), MicroRNAs of Epstein-Barr virus promote cell cycle progression and prevent apoptosis of primary human B cells. PLos Pathog 6: e1001063.

Seoud M (2012), Burden of human papillomavirus-related cervical cancer disease in the extended Middle East and north Africa: A comprehensive literature review. J Low Genit Tract 16: 106–120.

Sosa C, Benetucci J, et al (2001), Human herpesvirus 8 can be transmitted through blood in drug addicts. Medicine 61: 221–229.

Stewart SE, Eddy BE, Borgese N (1958), Neoplasms in mice inoculated with a tumor agent carried in tissue culture. J Natl Cancer Inst 20:1223–1243.

Sullivan CS, et al (2005), SV40-encoded miRNAs regulate viral gene express and reduce susceptibility to cytotoxic T cells. Nature 435:682–686.

Tao Q, Robertson KD (2003), Stealth technology: How Epstein-Barr virus utilizes DNA methylation to cloak itself from immune detection. Clin Immunol 109: 53–63.

Taylor GP, Matsuoka M (2005), Natural history of adult T-cell leukemia/ lymphoma and approaches to therapy. Oncogene 24:6047–6057.

Taylor GS, Blackbourn DJ (2011), Infectious agents in human cancer: Lessons in immunity and immunomodulation from gamma herpesvirus EBV ND kshv. Cancer Lett 305:263–275.

Temin HM, Mizutani S (1970), RNA-dependent DNA polymerase in virion of Rous sarcoma virus. Nature 226:1211–1213.

Terai M, Birk RD (2002), Identification and characterization of 3 novel genital human papillomavirus by overlapping polymerase chain reaction: candHPV89, candHPV90 and candHPV91. J Infect Dis 185:1794–1797.

Tolstov YI, Pastrana DV, et al. (2009), Human Merkel cell polyomavirus infection. I. MCV is a common human infection that can be detected by conformational capsid epitope immunoassay. Int J Cancer 125:1250–1256.

Trentin JJ, Yabe Y, Taylor G (1962), The quest for human cancer viruses. Science 137:835-841.:1579–1582.

Triboulet R, et al (2007), Suppression of microRNAs-silencing pathway by HIV-1 during virus replication. Science 315.

Vagin VV, et al (2006), A distinctive small RNA pathway silences selfish genetic elements in the germline. Science 313: 320–324.

Viswam S, et al (2012), Clinical outcome prediction by microRNAs in human cancer: A systematic review. JNCI 104:528–540.

Vlaar AP, et al (2011), Malignancies associated with chronic hepatitis C: Case report and review of the literature. Neth J Med 69:211–215.

Vogt PK, Rous P (1996), Homage and appraisal. FASEB J 10:1559–1562.

Vos P, Hagers R, et al (1995), AFLP: A new technique for DNA finger printing. Nucleic Acid Res 22:4407–4414.

Walboomers JMJM, Bosch FX, et al (1999), Human papillomavirus is a necessary cause of invasive cervical cancer worldwide. J Pathol 182:12–19.

Wenger RH, Kvietikova I, et al (1997), Hypoxia-inducible factor-1α is regulated at the post mRNA level. Kidney Int 51:560–563.

Wenger RH, Rolfs A, et al (1996), Nucleotide sequence, chromosomal assignment and mRNA expression of mouse hypoxia-inducible factor 1-α. Biochem Biophys Res Comm 223:54–59.

Wentzensen N, Sherman ME, et al (2009), Utility of methylation markers in cervical cancer early detection: Appraisal of the state-of-the-science. Gynecol Oncol 112: 293–299.

White MK, Gordon J, Khalili K (2013), The rapid expanding of human polyomavirus: Recent development in understanding their life-cycle and roles in human pathology. PLOS Pathog; 9: e10032006.

White MK, Pagano JS, Khalili K (2014), Viruses and human cancer: A long road to discovery of molecular paradigms. Clin Micro Rev 463:463–481.

Whiteside MA, Siegel EM, Unger ER (2008), Human papillomavirus and molecular considerations for cancer risk. Cancer 113: 2981–2994.

Yang Z, et al (2006), HENI recognizes 21-24 nt small RNA duplexes and deposits methyl group onto the 2' OH of the 3' terminal nucleotide. Nucleic acid Res 34:667–675.

Yeung ML et al (2008), Role for microRNAs, miR-93 and miR-130b, and tumor protein 53-induced nuclear protein I tumor suppressor in cell growth dysregulation by human T- cell lymphotrophic virus I. Cancer Res 68: 8976–8985.

Yin Q, et al (2008), Epstein-Barr virus-induced gene that modulates Epstein-Barr virus-regulated gene expression pathways. J Virol 82: 5295–5306.

Yoshida M, Miyoshi I, Hinuma Y (1982), Isolation and characterization of retrovirus human cell lines of human T-cell leukemia and its implication in the disease. PNAS USA 79:2031–2035.

Zur Hausen H (2006), *Infections causing human cancer.* New York: Wiley.

Zur Hausen H (1976), Condylomata acuminate and human genital cancer. Cancer Res 36:794.

Zur Hausen H (2009), Papillomaviruses in the causation of human cancers: A brief historical account. Virology 384: 260–265.

CHAPTER 2

ONCOVIRUSES AND THE IMMUNE SYSTEM: FIGHTING FOR SURVIVAL

Introduction

As stated earlier, oncoviruses have persistence, which enable them to develop into cancer. To prevent this, it is crucial for the host system to rapidly clear itself of all oncoviruses. Immunosuppressed patients like those with HIV infections are more prone to the development of oncoviral-associated cancers. This chapter will review the latest data on the immune responses in the recognition and clearance of oncoviruses. The activities of the immune system are divided into innate and adaptive immune responses; the innate response is non-specific and rapid, while the adaptive response is specific and requires activation. Both are required for effective clearance of oncoviruses.

2.1 Innate Immunity Response to Oncoviruses

2.1.1 Innate Response

The innate immune system is the first line of defence against invading pathogens. It includes physical barriers, humoral barriers, and cellular components. The epithelium forms the physical barrier against invading pathogens and is impermeable to most infectious agents. In addition, the epithelium secretes certain chemicals such as lysozyme and phospholipase to limit the invasion of pathogens. The humoral component of the innate immune system includes the complement system, coagulation system, antimicrobial peptides, chemokines, and cytokines. The invasion of a pathogen results in the epithelium cells releasing cytokines and chemokines that activate the immune cells present in the epithelium, such as Langerhans cells (LCs) found in the skin, and recruit these cells to clear the infection. The innate immune response depends on pattern recognition receptors (PRRs) to enable it to recognise Pathogen-Associated Molecular Patterns (PAMPs) of all class of pathogens. There are four families of receptors: Toll-like receptors (TLRs), nucleotide-binding oligomerisation domain (NOD)-like receptors (NLRs), retinoic acid inducible gene-I (RIG-1)-like receptors (RLRs), and the family of Pyhin. The PRRs are also involved in the sensing of endogenous ligands, referred to as Danger-Associated Molecular Patterns (DAMPs), released by cells in non-infectious conditions such as injury or cell death. These receptors are localised in the membrane (e.g., TLR) or in the cytosol (e.g., RLR, NLR). Their localisation is associated with the PAMP or DAMP they sense. TLRs specialise in virus recognition through sensing of nucleic acid located within the endosomes.

Aside from the TLRs, two families of cytosolic innate sensors have been described that are involved in virus recognition: RNA helicase and cytosolic DNA receptors, which include DAI, AIM2, and IFI16/p204. The cytoplasmic RNA helicase, RIG-1, and melanoma differentiation-associated genes 5 (MDA5) and DDXI/DDX21/DHX36 complex sense RNA or FNA viruses that replicate in the cytosol through an RNA intermediate. NLRs including NOD 1/2 are involved in bacterial recognition and the NLRP family members that sense various PAMP and DAMP. The binding of these receptors with their cognate ligands results in the recruitment of members of interleukin-1 receptor-associated kinase (IRAK) and tumour necrosis factor receptor-associated factor (TRAF) families, resulting in the activation of transcription factors such as NF-kB, interferon regulation factors (IRFs), and the mitogen-activated protein (MAP) kinase.

The activation of these pathways depends on various adaptors downstream of the different PRRs. Triggering PRRs results in the induction of proinflammatory cytokines, type 1 IFN, and chemokines that collectively participate in the innate and adaptive immune response. Type 1 IFN is one of the most important cytokines involved in viral infections. It is made up of thirteen subtypes. Type 1 IFN binds to α-IFN receptor 1 and 2 (IFNAR1 and -2), which leads to receptor dimerisation. In the classical pathway, receptor-bound Janus kinases (Jaks) and non-receptor tyrosine kinase 2 (Tyk2) are activated and cross-phosphorylate each other. The activated Jaks phosphorylate IFNAR1 and -2, which serves as Src homology 2 domain docking site transducers of activated transcription 1 and 2 (STAT1 and 2). These are phosphorylated by Jak1 and Tyk2. The phosphorylated STATs heterodimer interacts with IFN regulatory factor 9 (IRF9) to form the active transcriptional factor complex IFN-stimulated gene factor 3 (ISG3) (figure 1). This regulates expression of IFN-stimulated genes.

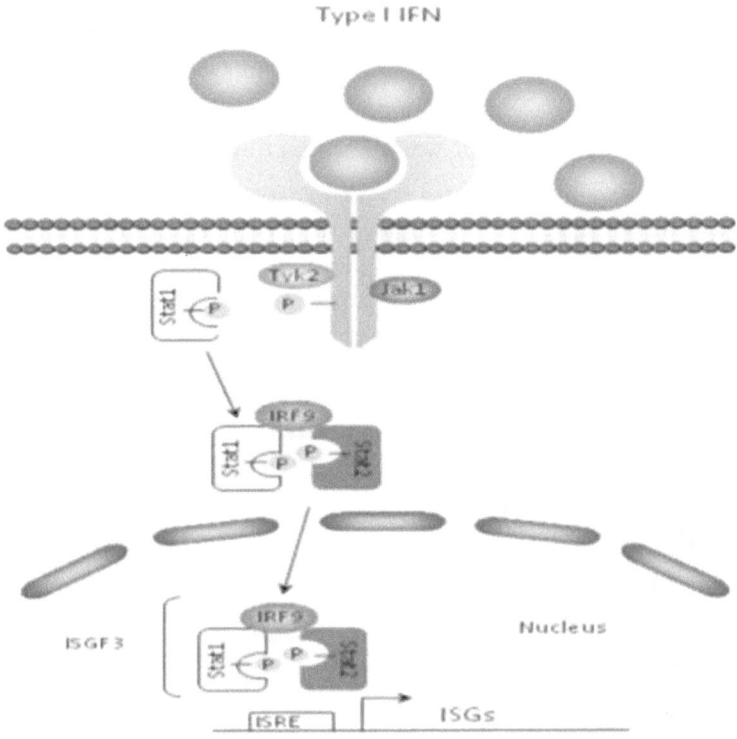

Figure 2.1: Type 1 IFN signal transduction (Source: Google image)

Type 1 IFNs are produced by most human cells during viral infections. Aside from their antiviral properties, IFN also have angiogenic, anti-proliferative, and immunostimulatory properties. Type 1 IFN modulates innate immune response by contributing to NK cell homeostasis and activation. In addition, they modulate adaptive immune response by inducing the phenotypical and functional maturation of immature DCs such as the upregulation of CD40, CD80, CD83, CD86, MHC Class 1 and II, and increased ability to stimulate T cell proliferation. Type 1 IFN allows the DCs to present endogenous antigen to CD8+ T cells, which allows for cross-priming of CD8+. This process is important for clearance of viral infection and cancer treatment.

The direct antiviral activities of type 1 IFN depends on three major IFN responses: the protein kinase K (PKR), Mx protein, and 2'5' oligoadenylate synthetase (2'5'OAS). The PKR is activated by dsRNA; caspases 3, 7, or 8; or polyanionic molecules. The main function of PKR after activation is to phosphorylate elf-2e, resulting in the blocking of protein translation. Furthermore, PKR also improves the induction of IFNβ and apoptosis induced by RLR as a result of measles viral infection. The Mx proteins are GTPase and involved in the binding to nucleocapsid of some viruses to alter their intercellular transport thereby interfering with the viral polymerase activities during viral transcription. The 2'5'OAS, on the other hand, activates RNAse due to the presence of dsRNA, in order to degrade viral and cellular RNAs, blocking the viral infection. Certain cytokines, IL-1β and IL-18, have been found to play an essential role in innate immunity. These cytokines are found in pro-form and then cleaved into active form by caspase-1. A multiprotein complex, referred to as inflammasome, is formed after the activation of NLRP-1 and -3 and triggering of DNA sensor AIMS. IL-1β is a proinflammatory cytokine involved in a number of inflammatory activities such as fever, stimulation of hepatic acute phase protein, hyperalgesia, and increase of bone marrow cell number. IL-18, on the other hand, is essential for priming of NK cells, which are critical for killing tumour cells lacking MHC Class 1 molecules.

In addition, these two cytokines also stimulate the innate immune response by activating neutrophils and macrophages, which engulf the pathogen and release reactive oxygen species (ROS). The response by IL-1 and IL-18 are essential for host defences against viruses or in the induction of innate immunity against tumours. Any breach or defect in their response results in enhancing the pathogenesis of viral infection or development of tumours.

2.1.2   Toll-Like Receptors in Oncoviral Infections

TLRs are analogous to the Toll receptor that was first described in the fruit fly *Drosophilia*. The Toll receptor plays a role in the development of dorso-ventral body axis in *Drosophilia*, acting as cell surface receptor for cytokine ligands. TLRs are type I transmembrane (TM) proteins made of extra-cellular leucine-rich receptors and are pattern recognition receptors expressed by various immune cells, such as macrophages, which recognise PAMP. The activation of the innate immune cells by TLRs initiate downstream signaling that leads to the upregulation of pro-inflammatory cytokines and chemokines, which recruits effector cells to the local site of infection. TLRs play an essential role in the induction of innate immunity. TLRs have been described as essential for adaptive immune responses. This is based on initial data which showed that molecules recognised as TLR ligands, such as LPS, enhance adaptive immune response.

However, there is no compelling evidence to show that there is adaptive immune failure in the absence of TLR signaling. Rather, robust adaptive immune responses, including antibody production, have been reported after infection in animals lacking TLR signaling. TLR signaling drives DC maturation, antigen presentation, and CD8+ cytotoxic T effector function. These are important for efficient antitumour immunity. So far, about a dozen mammalian TLRs have been described. Each TLR contains extracellular domains with leucin-rich repeats (LLR) motif that comprises a binding domain for recognition of their respective pathogens. The ligands for each TLR are different; for example, TLR3, TLR7, TLR8, and TLR9 recognise nucleic acid from bacteria and virus, while TLR2 recognises bacteria lipoprotein, lipotechoic acid, and fungal zymosans. TLR8 is required to sense HIV and activate inflammasome, while TLR7 senses HCV.

Aberrant TLR expression or chronic stimulation can alter homeostasis, which has negative regulation of antitumour immunity as shown by data which described increased immune suppression function, such as enhanced regulatory T-cell proliferation. Chronic activation of TLRs promoted the process of carcinogenesis through pro-inflammatory response and anti-apoptotic and pro-fibrogenic signals in the tumour microenvironment and in the tumour cells. TLR signaling, in addition to other inflammatory pathways, plays an important role in inflamed tumour microenvironment, resulting in inflammation-driven disease progression, as found in *Helicobacter pylori* associated gastric cancer.

In HPV-associated cervical cancer, for example, it has been suggested that TLRs play a significant role in the tumorigenesis process by modulating TLR expression and interfering with TLR signaling pathways, leading to persistent viral infection and carcinogenesis. A study by Zhang et al. to investigate the expression of TLR8 and its relationship with Bcl-2 (regulators of apoptosis and autophagy) and VEGF (potent angiogenic factor in tumour angiogenesis) in cervical cancer reported increased mRNA levels of TLR8 in cervical cancer cells as well as HeLa cells. In addition, there was positive correlation between the expression of TLR8, Bcl-2, and VEGF in cervical cancer. They therefore concluded that in patients with cervical cancer and HeLa cells, mRNA expression levels of TLR8 were upregulated, which was consistent with increased expression of Bcl-2 and VEGF.

This study shows that high expression of TLR8 may strongly correlate with carcinogenesis and tumour invasion through the inhibition of TLR8-positive immune cells to recognise tumour or viral antigens that can influence antitumour immune response. More studies are needed to understand the role of this protein in cervical cancer-associated tumorigenesis and whether it can be used as a novel therapeutic target for HPV-associated cervical cancer. This is true with EBV-associated cancers where overexpression of TLRs have been reported.

The naïve B cell is the target of EBV infection in vivo. The human B-cell expresses cell surface TLR1,-2, and -6 in addition to endosomal

TLR7,-9, and -10. Some studies showed that TLR6,-7,-9, and -10 are highly expressed more on memory B cell than on naïve B cell. Martin et al. found that upon contact of EBV to B cell, a number of cell genes such as TLR7 and myeloid differentiation factor 88 (MyD88), a key component in TLR signal transduction, are upregulated. Another study reported high expression of TLR3 in EBV-associated nasopharyngeal carcinoma through induction of inflammatory response.

These data support the suggestion that by hijacking the TLRs signaling pathway, oncoviruses could promote tumour growth. A novel approach of inhibiting viral-associated tumours is utilising the concept of TLR agonists. For example, in HPV infection, approximately 15 percent of women who have high-risk HPV infection are not able to mount any effective immune response against HPV, and persistence of high-risk HPV infection is one of the major factors in the development of cervical cancer. Slow clearance rate and lack of effective immune response mean HPV is escaping immune detection. This means therapeutic interventions are needed to treat high-risk HPV infections. HPV infects Langerhans cells, which are responsible for initiating an effective immune response against HPV.

However, LC exposed to HPV does not induce specific T cell immune response. Utilising TLR agonists would be effective in treating high-risk HPV infections. Fahey et al. suggested that in HPV16 infection, TLR7 and -8 are expressed in human LC; therefore, using imidazoquinolines would activate LC exposed to HPV16, resulting in the induction of an HPV16-specific cell-mediated immune response. This author suggested in a systematic review that 5 percent imiquimod plus IM IFN can reduce the incidence of HPV-associated cervical cancer. Imidazoquinolines are TLR7- or -8 agonists, making it a potent innate immune modulator. More studies are needed to develop novel TLRs agonists due to the increasing cases of viral-associated cancers which are now a global public health problem.

2.2   Inflammatory Response and the Oncogenesis Process

The other arm of the immune response is the adaptive immune response, which is a response of antigen-specific lymphocytes to antigen, including the development of immunological memory. Adaptive immune responses are generated by clonal selection (an important paradigm in immunology) of lymphocytes. The adaptive immune response is mediated by B and T lymphocytes, and data shows that adaptive immunity has some antitumorigenic effect. The major goal of any immune response is to eradicate pathogen, mainly through the mechanism of inflammation. The details of adaptive immune response in relationship to oncoviral infections have been dealt with in a number of publications. This section will review the role of inflammation in the pathogenesis of oncoviral infections.

Does an inflammatory response have a role in cancer development? The data now shows that inflammatory condition is ideal for the development of cancer. The inflammatory response is a fundamental immune mechanism that involves a number of molecular and cellular components, consisting of cytokines and chemokines that are released by proinflammatory cells. At the same time, some endogenous recruited components release anti-inhibitory mediators so that homeostasis is restored. The tools and strategies utilised by viruses to hijack the immune response is mostly associated with regulatory T-cells (Treg) that can inhibit inflammation and antiviral responses to other effector cells. Treg, therefore, appears to play a dual role in cancer pathogenesis. In liver cancer, it is known that hepatocarcinogenesis is promoted by facilitated cellular turnover, which is induced by chronic tissue damage and permanent cell regeneration as a result of chronic inflammation and sometimes after viral infections.

After HBV and HCV infection, the human hepatocytes express and present the viral antigen to CD4+ T- and CD+ T-cells, which result in the clearance of the virus by cytolytic and noncytolytic effector mechanisms. The CD8+ T cells play a central role in the inhibition of viral replication mediated by cytokine and direct killing of infected hepatocyte, while the CD4+ T cells activation may direct Th1 response with the secretion of IL2, TNFα, and IFN. Furthermore, the stimulation of CD4+ T cells may induce Th2 cells, which secrete IL4, IL5, and IL10. A third cell, Th17, has been described and associated with hepatic chronic inflammation following HBV and HCV infections. T17 are induced proinflammatory CD4+ T cells which secrete specific inflammatory cytokines such as IL17, IL21, Il22, IL6, and IL26. Further studies have shown that T17 exhibits either proinflammatory or protumour functions, with high proportion reported in advanced tumour stages.

A number of studies have reported that HCC patients exhibit specific inflammatory immune response, and HBV-HCV-associated microenvironment is colonised by infiltrated inflammatory cells such as macrophage, NK, B, and T lymphocytes. From these data, the inflammatory products can be associated with viral hepatocarcinogenesis. Others reported the presence of these inflammatory cytokines in the activation of NFkB pathways. In most patients, the tumour progresses despite the mounted immune response. This strongly suggests that HCC escapes from the immune response, probably as a result of the HCC-suppressive microenvironment which is associated with an essential role in tumour progression. In EBV-associated cancer, there is strong evidence linking EBV and nasopharyngeal carcinoma (NPC) and even Hodgkin's lymphoma (HL), and both are associated with EBV latency II profile.

In NPC development, EBV does play crucial role, but as to how and when, it is still not known. But we know that all NPC tumour cells bear the EBV monoclonal viral gene. In addition, EBV is present in neoplastic cells of patients with HL, which is characterised by the presence of malignant Hodgkin's and Reed Sternberg (HRS) cells. Leukocyte infiltration is linked to tumour development and cancer progression. A number of studies explained that lack of efficiency of immune effector cells plays a role in the cancer pathogenesis process. One important feature of NPC is the presence of massive lymphoid infiltrate in primary tumours, which might be supported by inflammatory cytokines produced by the malignant NPC cells. Studies have shown that most of the tumour-infiltrating leucocytes (TIL) are CD3+ T cells, but CD4+ and CD8+ T-cells are also found in different populations, depending on the tumour specimen. Other cells such as NK, DC, IL1α, and CXCL10 have been found in NPC cells. These findings support the argument that inflammatory responses are associated with tumour development and progression

2.3 Immune Evasion by Oncoviruses

Viral infection triggers an array of immune responses, involving an early host response through the activation of PAMP as well as orchestrating an adaptive immunity. To allow infection, replication, and persistence, most oncoviruses employ a number of strategies to evade host immune response. Viral immune evasion ranges from modulation of cytokines and chemo-attractant expression to alteration of antigen presentation and downregulation of IFN-pathways and adherence molecules. A number of studies have shown the mechanisms employed by various oncoviruses to evade the immune system. Below is a description of some of the mechanisms employed by oncoviruses to evade host defence mechanisms.

2.3.1 Interfering with Interferon

Innate immune sensing of viral infection leads to the production of Type 1 IFN, especially IFN-α and IFN-β, which are produced by cell types depending on the viral infection and promote an antiviral state in surrounding cells by the induction of IFN-stimulated expression. Oncoviruses develop mechanisms that inhibit IFN activities. In chronic HCV infection, Lee et al. found that excess production of IFNα receptor 2a (IFNαR2a) is associated with interference of IFNα. This is in conformity with the findings that elevated levels of soluble IFNα2Ra in serum and urine was associated with hairy cell leukaemia and adenocarcinoma, which resulted in high resistance of IFN therapy. Another mechanism is suppressing the function of TLR. A study by Hasan et al. reported that infection of human epithelial cells with HPV16 promotes the function of an inhibitory transcriptional complex containing NF-kBp50-p65 and ERα induced by E7 oncoprotein, which led to the downregulation of TLR9. Other mechanisms are alteration with pre-entry, post-entry, and pre-integration events in which IFNs play active roles. More studies are needed to elucidate the molecular and cellular mechanism of this interference.

2.3.2   Molecular Mimicry (MM)

MM is defined as the theoretical possibility that sequence similarities between pathogen and host peptide are sufficient to lead to cross-activation of auto reactive T and B cells by pathogen-derived peptides. This means molecular mimicry represents a shared immunologic epitope with a microbe and the host.

With viral infection, studies have shown that it is sometimes associated with the initiation or exacerbation of autoimmune disease. Although the mechanism underlying this is still unclear, it is believed that one of the mechanisms is viral determinate mimicking host antigen, resulting in the triggering of self-reactive T cell clones to destroy host tissue. Molecular mimicry is associated with the induction of autoimmune diseases. One good example is in rheumatic fever, where autoimmune disease can develop as a result of infection with group A beta-hemolytic streptococci infection. The mechanism underlying this has been described by Zabriskine and Cunningham. In another study, Zhao et al. described the role of molecular mimicry in HSV-1 infection. Their study showed that an epitope expressed by a coat protein of HSV-1 KOS strain was recognised by autoreactive T cells that targeted corneal antigens in a murine model of autocrine herpes stromal keratitis. The mutants HSV-1 viruses which did not express this epitope did not induce autoimmune disease. Thus, the researcher concluded that expression of molecular mimics can influence the development of autoimmune disease after viral infection. Data on molecular mimicry in oncoviral infection is scarce, however; few researchers have attempted to explore the role of molecular mimicry in relation to the pathogenesis of oncoviruses. It has been shown that Hepatitis B virus polymerase shared an immunologic epitope with myelin basic protein (MBP). When a viral peptide was injected into rabbits, some of the animals developed an experimental autoimmune encephalomyelitis (EAE)-like disease, developed antibodies to MBP, and had T cell reactivity. Four virus proteins similar to two human macrophage inflammatory protein (MIP) chemokines, IL-6 and interferon regulatory protein (IRF), are encoded by KSHV genome. Similarly, in HIV-1 transmission, studies have shown that vMIP-1 is similar to human MIP chemokines in its ability to inhibit replication of

HIV-1 strains which are dependent on the CCR co-receptor. This means that these genes might have a role in KSHV and HIV-1 interaction. Others have associated molecular mimicry to atherosclerosis. With our knowledge on molecular mimicry increasing, it is highly essential for us to explore their role in chronic infection, especially in relationship to oncoviruses.

2.3.3   MHC Downregulation

T-cell-mediated response is crucial in the defence against intracellular pathogens. The MHC class 1 molecule is a common target used by most oncoviruses to evade the immune system. The mechanism of MHC's action was described earlier. To evade the immune system, KSHV encodes K3 and K5 zinc finger membrane protein that removes the MHC class 1 molecule from the cell surface, thereby downregulating the class 1 molecule. In EBV infection, the virus enters a state of latency in the B lymphocyte. This is characterised by EBNA-1 expression that is involved in the maintenance of viral DNA episomes. The glycine-alanine repeat domain of EBNA-1 confers the virus with the ability to evade the immune system. This inhibits MHC class 1-restricted presentation of EBNA-1 epitopes linked in cis. When the EBV enters the replicative phase of infection, about eighty genes are expressed, which induces a strong immune response, but still the virus replicates. A number of studies have shown that this is due to the activities of several proteins which interfere with different stages of MHC class 1 and class II antigen presentation. These include inhibition of TAP by BNLF2a, which prevents peptide loading by MHC class 1, and inhibition of MHC class II antigen presentation by gp42/gH Ig L. These mechanisms confer EBV lytic proteins with the ability to interfere with CD4+ and CD8+ immune response, thereby allowing viral replication.

2.3.4   Generation of Escape Mutants

Oncoviruses have the ability to evade the immune system by rapidly mutating. Viruses with RNA genomes or RNA replicative intermediates utilise low-fidelity polymerase to generate mutants that are antigenically different, resulting in immune system evasion. In HBV infection, for example, replication takes place through reverse transcription of an RNA intermediate. The prospect of generating mutant viruses is high. Selective pressure to evade the host immune clearance readily selects out escape mutants. The cytotoxic T cells mediate the clearance of HBV from the liver and contribute to liver damage. Most of the acutely infected adults resolve all their clinical symptoms; however, mutation in the viral genes may result in lack of response to the viral antigens. With HCV, an RNA virus, the intrinsic infidelity of the HCV RNA polymerase generates many quasispecies that might be associated with immune evasion. In HTLV-1-associated HAM/TSP, the virus persists in the host despite a vigorous cellular and antibody response, which suggests that the virus has developed effective mechanisms to counteract the surveillance of the host's immune system. The open reading frame-1 (orf-1) has two products, P12 and P8, that increase the activity of NFAT. Mutation in the orf-1 gene results in the development of a mutant virus that is associated with immune evasion, viral replication, and persistence.

2.4 Advancing Oncovirus-Associated Cancer Immunotherapy with Checkpoint Immune Inhibitors

Immunotherapy is a promising therapeutic intervention in cancer cases, with immune checkpoint inhibitors gaining interest in the therapeutic development for cancer treatment. Immune checkpoints are inhibitory pathways directly connected into the immune system; they are crucial for maintaining self-tolerance and attenuating excessive immune reaction, which is important for maintaining homeostasis and minimising collateral tissue damage. Immune checkpoint inhibitors are also known as co-inhibitory molecules or co-stimulatory molecules expressed on T-cells; thus, immune checkpoints mediate either positive or negative signals that modify MHC-TCR signaling pathways. These signals each regulate T-cell survival, proliferation, differentiation, or responses to cognate antigens. Therefore, the net effect is dependent on the balance among the signals. T-cell activation requires co-stimulatory signals. If the antigen get connected with the co-stimulatory signals on APC, T-cells remain in a state of anergy. The co-inhibitory molecules induce T-cell dysfunction or apoptosis. When the immune system utilises this inhibitory pathway, it can attenuate excessive immune reactions and ensure self-tolerance. These functions include programme cell death protein-1 (PD-1), programme cell death-1 ligand1/2 (PD-L1/2), cytotoxic T lymphocytes antigen 4 (CTLA-4), lymphocyte-activation gene 3 (LAG-3), T-cell immunoglobulin mucin-3 (TIM-3), and B and T lymphocytes attenuator (BTLA). Tumour cells can bring together these suppressing effects as one of their immunoediting mechanisms. A number of studies have shown that immune checkpoint inhibitors with monoclonal antibodies promoted endogenous antitumour activities of immune cells. This section will review some of the checkpoint inhibitors available and the prospect of such inhibitors in oncoviral infection.

A number of immune checkpoint inhibitors have been developed and are in clinical trials. PD-1 and PD-Li inhibitors induce higher response rates across a wide range of tumours than other immunotherapy. They therefore provide a novel therapeutic intervention strategy in oncoviral infections. They have a lower rate of high-grade toxicities, mainly immune-mediated side effects. In addition, it is less labour-intensive to administer than some other types of immunotherapy. It does not require personal preparation for each patient, as do dendritic cell vaccines such as Provenge and chimeric antigen receptor (CAM) T-cell therapy, resulting in a longer duration of response.

Ipilimumab is a humanised IgG1 monoclonal antibody that inhibits CTLA-4. A number of clinical studies have been undertaken to evaluate its effect in patients with different malignancies, such as melanoma, non-Hodgkin lymphoma, renal cell carcinoma, and prostate cancer. In a phase I study to evaluate a single 3mg/kg dose of Ipilimumab in patients with metastatic hormone-refractory prostate cancer, it was shown that two (14 per cent) of the fourteen patients showed $\geq$50 per cent decline in prostate specific antigen. A patient developed a grade 3 rash which required systemic administration of corticosteroid. A phase II trial compared three doses: 0.3, 3.0, or 10 mg/kg administered every three weeks for a total of four doses. Eligible patients were allowed to receive reinduction or maintenance therapy. The overall response rate (ORR) in the 10mg/kg arm was superior to those in the other arms, but immune-related adverse events (irAEs) were higher in the 10mg/kg arm. The US FDA approved the use of Ipilimumab after phase III randomised controlled trials (RCTs) showed survival benefit.

Another agent is Pembrolizumab, a humanised IgG-4k antibody that blocks PD-1. A phase I dose escalation study evaluated three dose levels: 1, 3, and 10mg/kg, which were administered every two weeks in patients with multiple solid tumours. All dose levels were found safe, but the maximum tolerated dose was not identified. Clinical response was observed at all dose levels. In 2014, Pembrolizumab received approval for the treatment of patients with advanced melanoma by the FDA after reviewing the drug under its Accelerated Approval program.

Other immune checkpoint inhibitors being evaluated include BMS 936559 (for a number of cancers such as melanoma and ovarian cancer), MPDL3280A (as monotherapy for advance melanoma), and Nivolumab, for melanoma and non-small cell carcinoma of the lungs (NSCLC). There are no data on evaluating any immune checkpoint inhibitors in tumour virology, but take into consideration that most of these oncoviruses are known to use signaling pathways in the course of the pathogenesis; for example, in chronic HIV infection, molecules of the B7:CD28 family PD-1, CTLA-4, and their ligands play active and reversible roles in virus-specific T-cell exhaustion associated with HIV infection in humans and SIV models in macaques. CTLA-4 was found to be moderately overexpressed in CD4 population with progressive HIV infection, and its expression was inversely correlated with CD4 count. CTLA-4 was also reported to be strongly expressed in HIV-specific CD4 T cells at the time of acute HIV infection. A study reported that in HIV-infected subjects at different stages of HIV infection, CTLA-4 was upregulated on HIV-specific CD4 in all categories of HIV-infected individuals, with the exception of controllers who controlled viremia in the absence of ART therapy.

With PD-L1, its role in CTL exhaustion was shown initially in murine LCMV. A study by Barber et al. found that PA-1 was expressed on early effector CD8 T cells after infection with a LCMV strain that leads to both chronic infection and persistent viremia. It was reported that in HIV infection, high levels of PD-1 and its ligand PD-L1 and L2 are expressed which are found on hematopoietic and non-hematopoietic cells.

All these studies highlight the importance of these pathways in the HIV pathogenesis. This means inhibiting PD-L1 expression by a specific drug will be a novel strategy to manage chronic HIV infection and HIV-associated cancer. However, data shows that many cancer patients do not respond to therapeutic immune checkpoint intervention because of lack of tumour-infiltrating effector T-cells.

Therefore, the novel option in this regard is combinational intervention with vaccine. Cancer vaccine may prepare patients for treatment with checkpoint inhibitors by inducing effector T-cell infiltration into the tumour and checkpoint signals. Therefore, the combination of cancer vaccine and an immune checkpoint inhibitor may function simultaneously to induce more effective antitumour responses. However, others have proposed combinational therapy involving two or more different checkpoint inhibitors, such as PD-1 plus PD-L1, which is being tried by Astra Zeneca. The rationale for this is that while both therapies block the interaction between PD-1 and PD-L1, PD-1 inhibitors additionally block the interaction between PD-1 and PD-L2, another ligand of the receptor, while PD-L1 inhibitor additionally blocks the interaction between PD-L1 and B7.1 (also a ligand for CTLA-4). Analysis in preclinical models showed that this combination strategy has some synergy.

REFERENCES

Akira, S, Uematsu S, and Takeuchi, O (2006), Pathogen recognition and innate immunity. Cell 124:783–801.

Agathanggelou A, Niedobitek G, et al. (1995), Expression of immune regulatory molecules in Epstein-Barr virus-associated nasopharyngeal carcinomas with prominent lymphoid stroma evidence for a functional interaction between epithelial tumor cells and infiltrating lymphoid cells. The American Journal of Pathology 147:1152–1160.

Aggarwal BB, Shishodia S, et al. (2006), Inflammation and cancer: How hot is the link? Biochemical Pharmacology 72:1605–1621.

Albrecht B, D'Souza CD, et al. (2002), Activation of nuclear factor of activated T-cells by human lymphotropic virus type 1 accessory protein P12(I). J Virol 76:3493–3501.

Allen IC, et al. (2009), The NLRPs inflammasome mediates in vivo innate immunity to Influenza A virus through recognition of viral RNA. Immunity 30:556–565.

Ambrus JL, et al. (2003), Free interference in alpha/beta receptors in the circulation of patients with adenocarcinoma. Cancer 98:2730–2733.

Awasthi S, Sharma A, et al. (2005), A human T cell lymphotropic virus type 1 enhance of Myc transforming potential stabilizes Myc- TIP60 transcriptional interaction. Mol Cell Biol 25: 6178–6198.

Barber DL, Wherry EJ, et al. (2005), Restoring function in exhausted CD8 T cells during chronic viral infection. Nature 439:682–687.

Bogdan C (2000), The function of type I interferon in antimicrobial immunity. Curr Opin Immunol 12:419–424.

Brahwer JR, Tykodi SS, et al. (2012), Safety and activity of anti-PD-L1 antibody in patients with advanced cancer. NEJM 366:2455–2465.

Brunello MR, Rodriguez UA, Bonino F (1999), Hepatitis B virus mutants. Intervirology 42:69–80.

Budhu A, Wang XW (2006), The role of cytokines in hepatocellular carcinoma. J of Leukocyte Biol 80:1197–1213.

Busson P, Braham K, et al. (1987), Epstein-Barr virus-containing epithelial cells from nasopharyngeal carcinoma produce interleukin 1 alpha. PNAS USA 84:6262–6266.

Caramalho I, Lopes-Carvalho T, et al. (2013), Regulatory T cells selectively express toll-like receptors and are activated by lipopolysaccharide. J Exp Med 497:403–411.

Castana-Rodriguez N, Kaakoush NO, Mitchell HM (2014), Pattern recognition receptors and gastric cancers. Front Immunol 5:336.

Chaix J, et al. (2008), Cutting edge: Priming of NK cells by IL-18. J Immunol 181:1627–1631.

Chattergoon MA, Latanich R, et al. (2013), HIV and HCV activates the inflammasome in monocytes and macrophages via endosomal Toll-like receptors without induction of type 1 interferon. PLOS Pathog 10: e1004082.

Chen DS, Irving BA, Hodi F (2012), Molecular pathways: Next generation immunotherapy-inhibiting programmed death-ligand 1 and programmed death-1. Clinical Cancer Res 18:6580–6587.

Chen L, Flies DB (2013), Molecular mechanism of T cell co-stimulation and co-inhibition. Nat Rev Immunol 13:227–242.

Conroy H, Marshall NA, Mills KH (2008), TLR ligand suppression or enhancement of Treg cells? A double edged sword in immunity to tumors. Oncogene 27:168–180.

Cunningham MW (2000), Pathogenesis of group A streptococcal infection. Clin Microbiol Rev 13:420–511.

Diaz M et al. (1994), Structure of the human type I interferon gene cluster determined from a YAC clone contig. Genomics 22:540–546.

Dinnarello CA (1996), Biologic basis for interleukin-1 in disease. Blood 87:2095–2147.

Dong H, Zhu G, et al. (2004), B7-HI determines accumulation and deletion of intrahepatic CD8+ T lymphocytes. Immunity 20:327–336.

Epstein JE, Zhu J, et al. (2000), Potential roles of pathogen burden and molecular mimicry. Arteriosclerosis, thrombosis and vascular biology 20: 1417–1420

Fahey LM, Raff AB, et al. (2009), Reversal of human papillomavirus-specific T cell immune suppressor through TLR agonist treatment of Langerhans cells exposed to human papillomavirus type 16. J of Immunol 182:2919–2928.

Ferreira SH et al. (1988), Interleukin-1 beta as a potent hyperalgesic agent antagonized by a tripeptide analogue. Nature 334:698–700.

Flavell R A, Sanjabi S, et al. (2010), The polarization of immune cells in the tumour environment by TGFβ. *Nature Reviews Immunology* 10:554–567.

Franchini G, Laimore MD (2003), Human T-cell leukemia/lymphoma virus type 1 and 2 In: *Fields virology*. Philadelphia: LWW, 2071–106.

Fujinami RS, Oldstore MBA (1985), Amino acid and homology between encephalitogenic site of myelin basic protein and virus mechanism for autoimmunity. Science 230:1043–1045.

Fukata M, Abreu MT (2008), Role of toll-like receptors in gastrointestinal malignancies. Oncogenes 27:234–243.

Furness AJ, Vargas FA, et al. (2014), Impact of tumor microenvironment and Fc receptors on the activity of immunomodulatory antibodies. Trend in Immunology 35:290–298.

Gabay C et al. (2001), Production of IL-1 receptor antagonist by hepatocytes is regulated as an acute-phase protein in virus. Eur J Immunol 31:490–499.

Gay NJ, Ganghoff M (2007), Structure and function of Toll receptors and their ligands. Ann Rev Biochemistry 76: 141–165.

Goodbourn S, Didcock L, Randall RE (2000), Cell signaling, immune modulation, antiviral response and virus countermeasures, J Virol 81:12720–12729.

Hassan UA, Zannetti C, et al. (2013), The human papillomavirus type 16 E7 oncoprotein induces a transcriptional repressor complex on the Toll-like receptor 9 promoter. JEM 2:1369–1389.

Haybaeck J, Zeller N, et al. (2009), A lymphotoxin-driven pathway to hepatocellular carcinoma. Cancer Cells 16:295–308.

Hoos A, Ibrahim R, et al. (2010), Development of Ipilimumab: Commitment to a new paradigm for cancer immunotherapy. Seminar in Oncology 37:533–546.

Hoshida Y, Villanueva A, et al. (2008), Gene expression in fixed tissues and outcome in hepatocellular carcinoma. *The New England Journal of Medicine* 359:1995–2004.

Huang Y.-T., Sheen T.-S, et al. (1999), Profile of cytokine expression in nasopharyngeal carcinomas: A distinct expression of interleukin 1 in tumor and CD4+ T cells. Cancer Research 59:1599–1605.

Ito A, Kondo S, et al. (2015), Clinical development of immune checkpoint inhibitors. Biomed Res Int doi: 10.1155/2015/605478.

Ito T et al. (2001), Differential regulation of human blood dendritic cells subsets. J Immunol 166:2961–2969.

Ishida T, Ueda R (2006), CCR4 as a novel molecular target for immunotherapy of cancer. *Cancer Science* 97:1139–1146.

Kanodia SL, Fahey M, Kast WM (2007), Mechanism used by human papillomavirus to escape the host immune response. Curr Cancer Drug Targets 7:79–89.

Karin M, Greten FR (2005), NF-κB: Linking inflammation and immunity to cancer development and Progression. *Nature Reviews Immunology* 5:749–759.

Kaufman DE, Kavanagh DG, et al. (2007), Upregulation of CTLA-4 by HIV-specific CD4 (+) T cells correlates with disease progression and declines a reversible immune dysfunction. Nat Immunol 8:1246–1254.

Kawai T, Akira S (2010), The role of pattern recognition receptor in innate immunity. Nat Immunol 11:373–384.

Kim SJ, Ding W, et al. (2003), A conserved calcineurin-binding motif in human T lymphotropic virus type I p12 functions to modulate nuclear factor of activated T cell activation. J Biol Chem 278:15550–15557.

Kleponis J, Skelton R, Zhang L (2015), Fueling the engine and releasing the break: Combinational therapy of cancer vaccine and immune checkpoint inhibitors. Cancer Biol Med 12: 201–208.

Knolle PA, Thimme R (2014), Hepatic regulation and its involvement in viral hepatitis infection. Gastroenterology 146:1193–1207.

Knox P. G., Li Q.-X., et al. (1996), In vitro production of stable Epstein-Barr virus-positive epithelial cell clones which resemble the virus: Cell interaction observed in nasopharyngeal carcinoma. Virology 215:40–50.

Kruczek I, Wei S, et al. (2007), Cutting edge: Th17 and regulatory T cell dynamics and regulation by IL-2 in the tumor microenvironment. J of Immunol 178:6730–6733.

LaFleur DW et al. (2001), Interferon-kappa, a novel type I interferon expressed in human keratinocytes. J Biol Chem 5:39765–39771.

LeBon A et al. (2003), Cross-priming of CD8+ T cells stimulated by virus-induced type I interferon. Nat Immunol 4:1009–1015.

Lee Y-J, Zhang X, et al. (2014), Impaired HV clearance in HIV/HCV coinfected subjects treated with Peg IFN and RBV due to interference of IFN signaling by IFNαR2a. J Interferon Cytokine Res 34:26–34.

Li K, Lemon SM (2013), Innate immune response in hepatitis C virus infection. Semin Immunopathol 35: 53–72.

Li Z, Duan Y, et al. (2015), EBV-encoded RNA via TLR3 induces inflammation in nasopharyngeal carcinoma. Oncotarget 6: 1–13.

Lipson EJ, Drake CG (2011), Ipilimumab: An anti-CTLA-4 antibody for metastatic melanoma. Clinical Cancer Res 17:6958–6962.

Locarnini S, Zoulin F (2010), Molecular genetics of HBV infection. Antiviral Ther 15 (Suppl 3): 3–14.

Lopes AR, Kellan P, et al. (2008), Bim-mediated deletion of antigen-specific CD8+ T cells in patients unable to control HBV infection. J Clin Invest 118:1835–1845.

Maker AV, Phan GQ, et al. (2005), Tumor regression and autoimmunity in patients treated with cytotoxic 4 blockades and interleukin 2: A phase I/II study. Annals of Surg Oncol 12:1005–1016.

Martin HJ, Lee JM, et al. (2007), Manipulation of the Toll-like receptor 7 signaling pathway by Epstein-Barr virus. J Virol 81:9748–9758.

McAllister CS, Sameul CE (2009), The RNA-activated protein kinase enhances the induction of interferon-beta and apoptosis mediated by cytoplasmic RNA sensors. J Biol Chem 284:1644–1651.

Montaya M, et al. (2002), Type I interferon produced by dendritic cells promote their phenotypic and functional activation. Blood 99:3263–3271.

Niedobitek G, Young L. S. et al. (1992), Expression of Epstein-Barr virus genes and of lymphocyte activation molecules in undifferentiated nasopharyngeal carcinomas. The American Journal of Pathology 140:879–887.

Nielsen BW, et al. (1994), Macrophages as producers of chemotactic proinflammatory cytokines. Immuno Ser 60:131–142.

Nishikawa H, Sakaguchi S (2014), Regulatory T cells in cancer immunotherapy. *Current Opinion in Immunology* 27:1–7.

Novick D, et al. (1992), Soluble interferon-alpha receptor molecules are present in body fluids. FEBB Lett 314:445–448.

Numasaki M, Fukushi J-I, et al. (2003), Interleukin-17 promotes angiogenesis and tumor growth. Blood 101:2620–2627.

O'Connor JP Manigrasso M, et al. (2014), Fracture healing and lipid mediators. *BoneKEy Reports* 2014; 3:517.

Page DB, Postow MA, et al. (2014), Immune modulation in cancer antibodies. Ann Rev Med 65:185–202.

Pardoll DM (2012), The blockade of immune checkpoints in cancer immunotherapy. Nat Rev Cancer 12:252–264.

Pathmanathan R, Prasad U, et al. (1995), Clonal proliferations of cells infected with Epstein-Barr virus in preinvasive lesions related to nasopharyngeal carcinoma. The New England Journal of Medicine 333:693–698.

Patnaik A, Kang S, et al. (2012), Phase I study of MK-3475 (anti PD-1 monoclonal antibody) in patients with advance solid tumors. Proceedings of the ASCO annual meeting, May 2012.

Phan GQ, Yang JC, et al. (2003), Cancer regression and autoimmunity induced by cytotoxic T lymphocyte-associated antigen and blockade in patients with metastatic melanoma. PNAS of USA 100:8372–8377.

Pise-Masison CA, de Castro-Amarante MF, et al. (2014), Co-dependence of HTLV-1 p12 and p8 function in virus persistence. PLOS Pathog 10: e1004454.

Platanias LC (2005), Mechanisms of type-I and type-II-interferon-mediated signaling. Nat Rev Immunol 5:375–386.

Ramzan MSN, Thĕlu MA, et al. (2012), Intra-hepatic lymphocytes in HCV-related cirrhosis and hepatocellular carcinoma. Gut 2012.

Reherman B, Nascimbeni M (2005), Immunology of hepatitis B virus and hepatitis C virus infection. Nat Rev Immunol 5:215–229.

Ressing ME, Hoist D, et al. (2008), Epstein-Barr evasion of CD8(+) and CD4(+) T cell immunity via concerted action of multiple genes products. Semin Cancer Biol 18: 397–408.

Ruvola VR, et al. (2008), PKR regulates B56 (alpha)-mediated BCL 2 phosphotase activity in acute lymphoblastic leukemia-derived REH cells. J Biol Chem 283:35474–35484.

Schreiber RD, Old LJ, Smyth MJ (2011), Cancer immunoediting: Integrating immunity's roles in cancer suppression and promotion. Science 331: 1565–1570.

Stanley MA, Pett MR, Coleman N (2007), HPV: From infection to cancer. Biochem Soc Trans 35:1456–1460.

Stassen M, et al. (2000), Murine bone marrow-derived mast cells as potent producers of IL-9 costimulatory function of IL-10 and kit ligand in the presence of IL-1. J Immunol 164:5349–5355.

Swann JB, et al. (2007), Type 1 IFN contributes to NK cell homeostasis, activation, and antitumor function. J Immunol 178:7540–7549.

Takeda K, Akira S (2005), Toll-like receptor in innate immunity. Int Immunol 17:1–14.

Teichmann M, Meyer B, Beck A (2005), Expression of the interferon-inducible chemokine IP-10 (CXCL10), a chemokine with proposed anti-neoplastic functions, in Hodgkin lymphoma and nasopharyngeal carcinoma. The Journal of Pathology 20668–75.

Thomas E, Gonzale VB, et al. (2011), HCV infection induces a unique hepatic innate immune response associated with robust product of type III interferon. Gastroent 142:978–988

Topalian SL, Drake CG, Pardoll DM (2012), Targeting the PD-1/B7-HI (PD-L1) pathway to activate anti-tumor immunity. Curr Opin in Immunology 24:207–212.

Wolchok JD, Neyns B, et al. (2010), Ipilimumab monotherapy in patients with advanced melanoma: A randomized, double-blind, multicenter, phase 2, dose-ranging study. The Lancet Oncology 11:155–164.

Wucherpfennig KW, Strominger JL (1995), Molecular mimicry in T-cell-mediated autoimmunity: Viral peptides activate T-cell specific myelin basic protein. Cell 80:695–705.

Wu W, Li J, et al. (2010), Circulating Th17 cells frequency is associated with the disease progression in HBV infected patients. J of Gastroenterol 25:750–757.

Yang J, Zheng J-X, et al. (2012), Hepatocellular carcinoma and macrophage interaction induced tumor immunosuppression via Treg requires TLR4 signaling. World J of Gastroenterology 18: 2938–2947.

Zabriskine JB (1982), Rheumatic fever: A streptococcal induced autoimmune disease. Pediatric Ann 11: 383–396.

Zampronio AR, et al. (1994), Interluekin-8 induces fever by a prostaglandin-independent mechanism. Am J Physiol 266:R1620–R1624.

Zhang J.-P, Yan J, et al. (2009), Increased intratumoral IL-17-producing cells correlate with poor survival in hepatocellular carcinoma patients. *Journal of Hepatology* 50:980–989.

Zhang Y, Yang H, et al. (2014), The expression of Toll-like receptors and its relationship with VEGF and Bcl-2 in cervical cancer. Int J of Medical Sciences 11:608–613.

Zhao Z-Shan, Granucci F, et al. (1998), Molecular mimicry by Herpes simplex virus-type 1: Autoimmune disease after viral infection. Science 279: 1344–1347.

Zuo J, Currin A, et al. (2009), The Epstein-Barr G-protein-coupled receptor contributes to immune evasion by targeting MHC Class 1 molecule for degradation. Plos Pathog 5: e1000255.

CHAPTER 3

VIRULENCE FACTORS AND ENTRY MECHANISM OF HTLV

Introduction

Human T-cell lymphoma virus (HTLV) or T-lymphotropic virus type 1 is a member of the deltaretrovirus family, which includes the simian T-lymphotropic virus type 1 (STLV-1) and bovine leukaemia virus (BLV). There are four known strains of HTLV: HTLV-1, HTLV-2, HTLV-3, and HTLV-4. HTLV-1 and -2 are prevalent around the globe, while HTLV-3 and -4 have been reported only in central Africa. HTLV-1 and STLV-1 are highly related to STLV-2 and -3, respectively. It is believed that HTLVs evolved through interspecies transmission between monkeys and humans. Genetic variation among HTLV-1 strains is reported to be less than 8 per cent, while HTLV-1 and HTLV-2 shows 70 per cent nucleotide homology. HTLV-1 is the first retrovirus to be described, and it is estimated that about 20 million people are infected with it worldwide. HTLV-1 is the direct cause of Adult T cell leukaemia/ lymphoma (ATLL) and also causes many other chronic inflammatory diseases, such as myelopathy/tropical spastic paraparesis (HAM/TSP), chronic renal failure, and monoclonal gammopathy due to the induction of immunodeficiency. The virus mainly infects CD4+ T lymphocytes without evidence of neuronal infection. HTLV-2 is rarely pathogenic and is associated with sporadic neurological disorder. No disease is associated with HTLV-3 and HTLV-4.

Epidemiologically, the virus is endemic in southwestern Japan, central Africa, the Caribbean Islands, and Australia (among the Aborigines). Available data shows that the virus prevalence has some ethnic distribution. In these endemic areas, the seroprevalence rate ranges from 0.1 to 30 per cent. After a long period of latency, an estimated 5 per cent of people infected with HTLV-1 will develop ATL. HTLV-2 is prevalent among IV drug users in the US, Europe, South America, and Southeast Asia. HTLV-1 is transmitted through three main routes: 1. Breastfeeding from mother-to-child. 2. Sexual contact from males to females. 3. Needle sharing mediated by exposure to contaminated blood. The mother-to-child transmission is the predominant mode of transmission. This chapter will review viral genes associated with HTLV pathogenesis.

3.1 Genomic Description of HTLV and Infectivity

The HTLV-1 is a single-stranded diploid RNA virus with a proviral genome 9,030–9,040 nucleotides long, containing two flanking long terminal repeat (LTR) sequences. The LTRs are made up of three components: a unique 3' (U3) region, a repeated (U) region, and a unique 5' region (U5). The HTLV-1 genome (figure 1) is packed in the viral core with the viral nucleocapsid protein (NC, p15), which is surrounded by capsid (CA, p24) and matrix (MA, p19) proteins. The virus encodes the structural protein Gag (NC, CA, and MA) and Env and the enzymes RT, RNase H (RH), integrase, and protease. The env gene encodes the surface unit (SU) gp46 and transmembrane unit (TM) gp 21 proteins. The pX region also encodes the accessory proteins as well as regulatory proteins Tax and Rex.

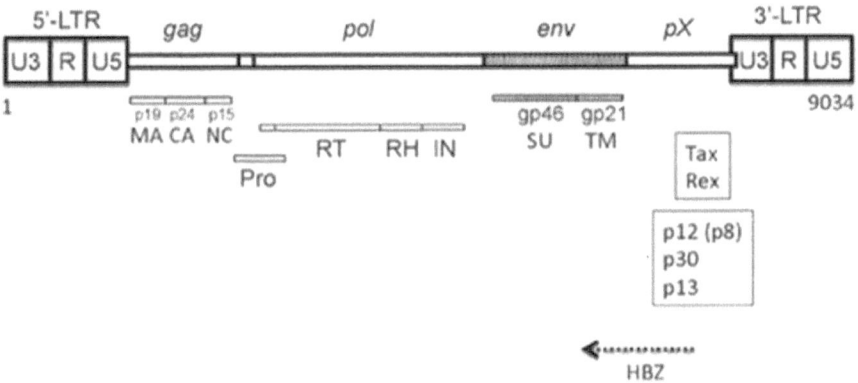

Figure 3.1: Structure of HTLV-1 genome: The gag gene encodes the matrix (MA), capsid (CA), and nucleocapsid (NC) proteins. The pol gene encodes reverse transcriptase (RT), RNase H (RH), and integrase (IN). The pX region encodes p13, p12, and p30. HBZ is encoded by the antisense frame of the provirus.

The accessory proteins are spliced alternatively and translated into different initiation sites. These genes encode novel proteins such as p12/p8, p30, and p13. The anti-sense region corresponds to the px region. The env gene encodes HBZ. The roles of these proteins will be discussed in details. The envelope protein of the virus interacts with glucose transporter GLUT-1, heparin sulfate proteoglycans (HSPGS), and neuropilin-1 (NRP1) to help the virus enter the cells. Productive infected HTLV-1 cells establish viral synapse through cell-to-cell contact with uninfected T cells mediated by interaction between ICAM-1 and LFA-1 adhesion molecule. The virological synapse mediates the accumulation and spread of the core complex of the virus and genome to uninfected cells.

3.2   Tax Protein

Tax, a transactivator/oncoprotein, is a regulatory protein that immortalises human CD4+ memory T cells, but when expressed alone, they rarely transform T-cells. This means there should be a co-factor which helps facilitate transformation. A study found that Tax proteins displayed differential ability to immortalise human CD4+Foxp3 T cell with characteristic expression of CTLA-4 and GITR. Tax impacts quite a number of cellular processes. All these activities have been implicated in the development of HTLV-1. These cellular factors include CREB and CREM proteins, NF-kB p50, p65, C-Rel proteins, and SRF. In addition to these transcription factors, Tax also binds to NK-kB p105, p100, and IKR proteins, which are inhibitors of NF-kB. The Tax binding domain (figure 2) is the IKB proteins which were identified as ankyrin motifs.

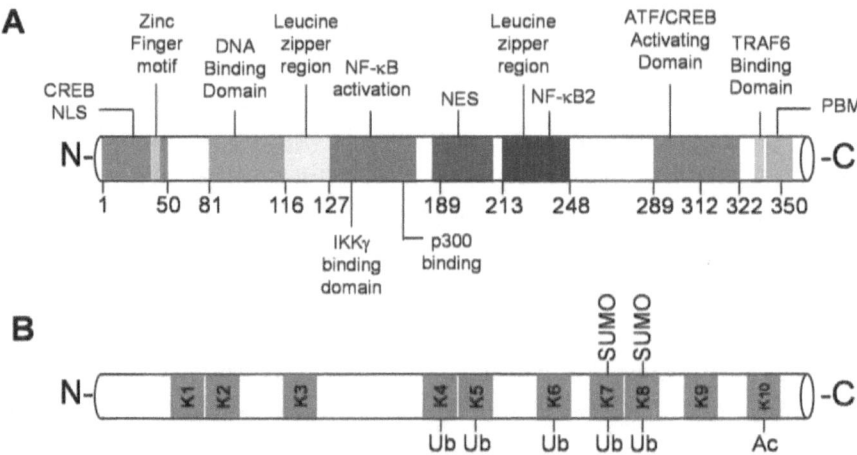

Figure 3.2A. Schematic representation of Tax protein-protein interaction domains. 2B. Schematic representation of known ubiquitination, SUMOylation, and acetylation of Tax (Source: Lavorgna and Harhaj, 2014)

The ankyrin motif are also found in other proteins involved in cell cycle control and tissue differentiation. This suggests that there is possible interaction between these proteins and Tax.

Tax is concentrated in different compartment of the cell, including cytosol, nucleus, Golgi apparatus, and endoplasmic reticulum (ER), which helps the virus to replicate and persist. Tax activates the transcription factor NF-kB in the cytoplasm and cis-Golgi, while also regulating viral gene expression in the nucleus. To coordinate these highly essential functions, Tax forcefully moves between different sub cellular compartments through nuclear localisation sequences (NLSs) and nuclear export sequences

(NESs). Tax localisation can be influenced by specific stimuli like genotoxic stress, which triggers Tax nuclear export. Tax can be secreted into the extracellular compartment, where cell-free Tax can play a role in inflammation and pathogenesis. The large number of cellular proteins and dynamic localisation patterns suggests that Tax has an effect on cell proliferation, survival, and pathogenesis. Tax perturbs signaling pathways leading to upregulation of host cell factors. Amongst these is the actin-binding protein Fascin, a major marker of several types of cancer. The role of Fascin has been elucidated in a study by Mohr et al. They identified a triple mode of transcriptional induction of Fascin which requires NF-kB-dependent promoter activation, a Tax-responsive region in the Fascin promoter, and a promoter-independent inhibitor of PP2. From this, they concluded that Tax regulated Fascin by a multitude of signals.

The NK-kB pathway is also a key cellular target. Tax interacts with specific components of NF-kB pathways to drive proliferation, survival, and transformation of HTLV-1-infected T cells. NF-kB is a family of transcription factors that regulates diverse functions, such as cell cycle, apoptosis, inflammation, and development of lymphoid organogenesis. It also causes the activation of the canonical NF-kB pathway. Furthermore, Tax activates host E3 ligase (TRAF6) to help stabilise MCL-1 and reduce the number of cell deaths triggered by genotoxic stress agents and chemotherapy drugs.

Tax inactivates a number of deubiquitinating enzymes DUBS that oppose NF-kB activation. Although Tax disturbs the host ubiquitin–proteasome pathway for NF-kB activation and cell transformation, there is still a lot to be done to understand the precise mechanisms. We have no clue as to the identity of the K63-U6-specific E3 or E4 or the potential roles of the E2s or E3s. As a result, mass-spectrometry-based proteomic screening or yeast two-based hybrid screening has a role in identifying ubiquitin–proteasome components that interact with Tax. In addition, the screening of E2 and E3 enzyme siRNA libraries might identify key component that regulate Tax function, stability, and trafficking. Elucidating the complex interplay between Tax and the host ubiquitin-proteasome mechanism might go a long way to identify drug targets against Tax, which will be useful in HTLV-1-associated diseases.

3.3 The HBZ Gene

The HTLV-1 bZIP factor (HBZ), just like Tax, is found in the pX region, and it is encoded in the plus-and-minus tail of the pX region. It has a number of functions on the T-cell signaling pathway and associated with HTLV-1 pathogenesis. The HBZ gene (figure 2) has two transcripts: a spliced form (sHBZ) and an unspliced form (usHBZ). The protein of the spliced and unspliced HBZ transcripts does not have TATA and contains initiators and downstream promoter elements. The spliced and unspliced HBZ genes are translated into a polypeptide of 206 and 209 amino acids, respectively. Both forms are made of three domains: N-terminal activation domain (AD), central domain (CD), and basic ZP domain (bZIP) in the C-terminal. There are two LXXLL-like motifs found within the N-terminal AD domain of HBZ.

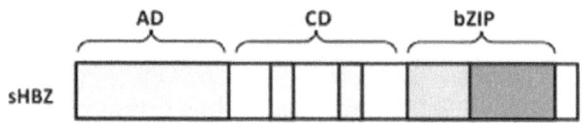

Domain	Function
AD	Bind with 26S proteasome
	Interact with CBP/p300
	Activate TGF-β signaling
CD	Induce nuclear localization
bZIP	Interact and inhibit c-Jun, JunB, CREB, CREB2
	Inhibit 5'LTR activation
AD+bZIP	Bind with p65, inhibit classical NF-kB pathway
	Ativate JunD
	Enhance hTERT promoter activity

Figure 3.3: Schematic representation of HBZ domain

Both are important for the binding of p300/CBP. The HBZ protein is localised in the nucleus in a speckled manner. Three nuclear localisation signals (NSLs) were identified, with two regions in the CD of HBZ and a basic region of the bZIP domain. The difference between the two transcripts are seven amino acids in their N-terminal, which causes significant difference in the two proteins. The half-life of usHBZ protein is much shorter than that of sHBZ in ATL. This is attributed to the fact the sHBZ gene is more dominant than suHBZ.

A number of studies have characterised the role of HBZ in HTLV-1-associated pathogenesis. In one study, Mitagami et al. found that HBZ transgenic (HBX-Tg) mice that express HBX-CD4+ T cells developed systemic inflammatory disease, cellular immunodeficiency, and T-cell lymphomas. This suggests that HBZ plays an important role in HTLV-1-mediated pathogenesis. In HBZ-Tg, the numbers of CD4+ CD25+ T cells and effector/memory CD4+ T-cells were increased, as in ATL cases.

In another study, Mitobe et al. reported that the transduction of mouse T-cells with specific mutants of HBZ that distinguished between its RNA and protein activity resulted in differential effects on T-cell proliferation and survival. The HBZ RNA increased cell numbers by attenuating apoptosis BZ, while HBZ protein induced apoptosis. However, both HBZ RNA and protein promoted S-phase entry of T-cells. They identified a HBZ 50bp coding sequence that was essential for RNA-mediated cell survival. When they profiled T-cells expressing wild-type HBZ, or protein, they found that HBZ RNA was associated with genes that promoted cell cycle proliferation and survival. HBZ protein was, however, more closely associated with immunological properties of T-cells. They reported on survivin, which inhibits apoptosis. Inhibition of survivin led to impaired proliferation of several ATL cells. This means that HBZ is essential in the pathogenesis of ATL.

Furthermore, HTLV-1 infection is associated with chronic inflammation in the CNS, skin, and lungs. Because HTLV-1 directly infects CD4+ T cells, it should modulate the host immune response, not only via viral antigen stimulation but also via CD4+ T cell-mediated immune deregulation.

Earlier, it was reported that Foxp3+ CD4+ T cells are increased in HTLV-1 infection. One of the central questions asked in relation to HTLV-1 pathogenesis is why HTLV-1 induces inflammation despite the increase in Foxp3+ cells. A study found that most increases in Foxp3+ cells in HBZ-Tg mice or HAM/TSP patients were not naturally derived Treg cells but rather induced Treg cells since iTreg cells could serve as source of proinflammatory CD4+. Therefore, Yamamoto-Taguchi et al. concluded that HTLV-1 causes abnormal CD4+ T cell differentiation by expressing HBZ, which is believed to play a crucial role in chronic inflammation associated with HTLV-1. This study further highlights the role of HBZ in the pathogenesis of HTLV-1-associated diseases. HBZ can therefore serve as a novel target for therapeutic interventions.

3.4   Rex Protein

HTLV Rex is a trans-acting regulatory protein responsible for the nuclear export of unspliced gag/pol and incompletely spliced env mRNAs into the cytoplasm. It binds to the Rex responsive element (RxRE) found on viral mRNAs and to the chromosome maintenance region 1 (CRM1) cellular export factor. Mutational analysis of HTLV-1 and HTLV-2 showed that Rex has several domains within the protein necessary for its function. These include the arginine-rich N-terminal RNA binding domain (RBD) that overlaps with the nuclear export signal (NES), and two regions that flank the NES which have been shown to be important for Rex-Rex multimerisation. A novel carboxyl terminal domain containing key phosphorylation sites essential for function has been described for HTLV-2 Rex. Rex function includes regulating the cytoplasmic levels of genomic RNA and expression of the structural as well as enzymatic gene products that are critical for production of virus progeny. It has therefore been suggested that Rex plays an essential role in the transition from early to late stage of HTLV-1 infection and is required for efficient spread of the virus. It has also been suggested that Rex's modulation may determine whether a productively infected cell becomes latent. Therefore, HTLV-1 absolutely requires Rex for efficient persistent infection in vivo.

3.5   Env Protein

Like other retroviruses, HTLV-1 virion is surrounded by a proteo-lipid envelope made of protein with transmembrane (TM) and surface (SU) subunits. The most important factor in HTLV-1 pathogenesis is the product of the viral env gene (figure 4), which is cleaved by cellular protease into non-covalently linked TM subunit gp21 and extracellular subunit gp46, with the latter made of an N-terminal receptor-binding domain (RBD), a mid-proline-rich region (PRR), and a C-terminal domain.

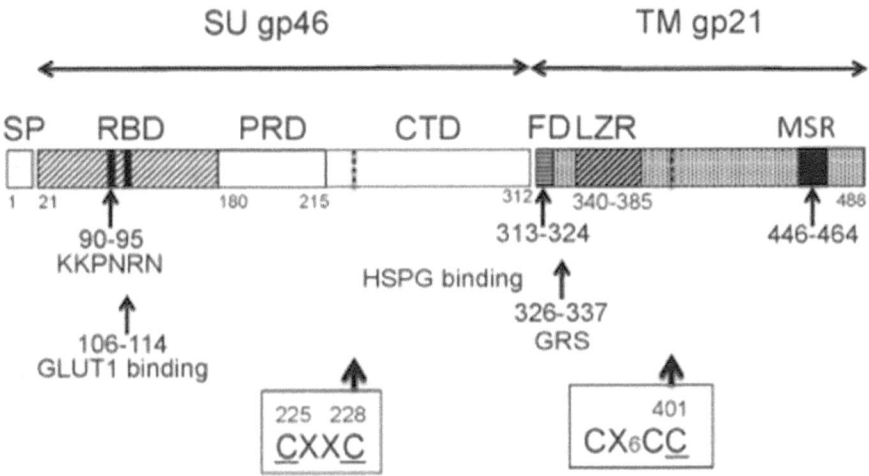

Figure 3.4: The structure of Env protein (Source: Hoshino, 2012)

HTLV-1 env gene is encoded by 488 amino acid precursor protein, which generates a 62kDa protein (gp62) after the addition of five N-glycan chains. Four are in the SU. The gp62 is cleaved at a trypsin-like proteolytic site spanning residues 309-312 into gp46 and gp21 subunit. The SU is entirely extracellular and remains linked to the virus through binding to TM, which is embedded in the viral envelope.

3.6   Mechanism of HTLV-1 Entry

Just like all retroviruses, HTLV-1 entry into target cells involves interaction between the viral envelope glycoprotein, SU, TM, and host cell receptors.

The SU protein is involved in receptor recognition, while the TM protein triggers the fusion of viral and cellular membranes, thereby allowing entry of viral particles. With some retroviruses, such as ecotropic murine leukaemia virus, a single molecule is enough for attachment and entry, while with others, such as HIV, multiple molecules are required. Our current understanding of HTLV-1 entry is based on data from several researchers which showed that entry into host cells is multireceptor dependent, involving glucose transporter I (GLUT I), neuropilin-I (NRP-I), and heparan sulfate proteoglycans (HSPG). GLUT-I specifically binds to a truncated soluble form of HTLV-I and HTLV-2 SU proteins. Studies showed that overexpression of GLUT-I in an HTLV-1-resistant cell line was associated with increased HTLV-1 titer. The HSPG is a glycosamine-glycine consisting of a core protein and heparan sulfate (HS) polysaccharide chains. Studies showed that enzymatic reduction of the cell surface levels of HSPG led to reduction of both the binding of soluble HTLV-1 SU and the titer of HTLV-1 Env pseudotype viruses in non T-cell lines. Jones et al. showed that HSPG is also involved in the binding and entry of HTLV-1 into CD4+ T cells. The NRP-1 is a cell surface protein known to function as co-receptor for some heparin-binding pro-angiogenic cytokines, principally members of the vascular endothelial growth factor (VEGF) family, and class 3 semaphorins. Studies have shown that NRP-1 binds HTLV-1 SU and is necessary for efficient HTLV-1 entry. One study also showed that NRP-1, GLUT1, and HTLV-1 SU form a stable tripartite complex when co-expressed in cells.

HTLV-1 entry is through viral attachment and virus/cell fusion. The initial attachment is mediated by the interaction between the C-terminal domain of the virus gp46 and HS moieties of cell surface HSPG, of activated CD4+ T cells, or DC. It has been found that the length of HS plays a role in the susceptibility of cells to HTLV-1 entry, with shorter chains facilitating closer attachment of the virus to the target cells. This leads to viral binding to NRP-1. The viral gp46 RBD domain contains KPxR motif, which has been identified as an NRP-1 binding site. The NRP-1 b-domain binds to VEGF and is required for HTLV-1 entry. HTLV-1 utilises HSPG and NRP-1 complex through molecular mimicry of VEGF. After stable binding of HTLV-1 to HSPG and NRP-1, GLUT-1 binding sites are exposed. This might be due to conformational changes in gp46. The HTLV-1 binds to GLUT-1 at its large extracellular loop (ECL1). The amino acid residues found in the RBD of gp46 that are essential for GLUT-1 binding are distinct from the binding sites of NRP-1 and HSPGs, indicating that a multireceptor complex is formed.

Antibodies to GLUT-1 alter HTLV-1 fusion and infection but not binding to CD4+ T cells. This means GLUT-1 is important for fusion. For cell-to-cell transmission to be effected, an infected cell must come in contact with an uninfected cell; this forms a virological synapse (VS). In a study, Igakura et al. reported that infected T-cells formed conjugate with neighbouring uninfected cells, resulting in the polarisation of the microtubule organisation centre (MTOC) to the point of contact. This polarisation of MTOC is promoted by the interaction of intercellular adhesion molecules -1 or -3 (ICAM-1, ICAM-3) or the vascular cell adhesion molecule-1 (VCAM-1) on infected cells with β-integrins, such as lymphocyte function associated antigen-1 (LFA-1) or uninfected cells. The HTLV-1 genome, Env, and Gag protein accumulate at the point of contact and then egress into the VS and interact with receptor on the conjugated, uninfected cell. Studies have shown that NRP-1 and GLUT-1 co-localise at the point of contact and presumably promote the formation of the VS, which is also facilitated by HTLV-1 Tax. Although little is known about this mechanism, it is believed that it occurs through endocytosis of the virus, as found in HIV.

REFERENCES

Albrecht B, Collins ND, et al. (2000), Human T-lymphotropic virus type 1 open reading fram I p12 (I) is required for efficient viral infectivity in primary lymphocytes. J Virol 74:9828–9835.

Araujo A, Hall WW (2004), Human T-lymphotropic virus type II and neurological disease. Ann Neurol 56: 10–19.

Arnold J, Yamamoto B, et al. (2006), Enhancement of infectivity and persistence in vivo by HBZ, a natural antisense coded protein of HTLV-1. Blood 107:3970–3982.

Chan AJ, Rosenblatt JD, et al. (1985), Identification of the genome responsible for human T-cell leukemia virus transcriptional regulation. Nature 318:571–574.

Chen L, Liu D, et al. (2015), Foxp3-dependent transformation of human primary CD4+ T lymphocytes by the retroviral protein Tax. Biochem Biophys Res Commun 466:523–529.

Coskun AK, Sutton RE (2005), Expression of glucose transporter I confers susceptibility to human T-cell leukemia virus envelope-mediated fusion. J Virol 79:4150–4158.

Feuer G, Green PL (2005), Comparative biology of human T-cell lymphotropic virus type 1 (HTLV-1) and HTLV-2. Oncogene 24:5996–6004.

Fujisawa J, Seiki M, et al. (1985), Functional activation of the long terminal repeat of human T- cell leukemia virus type 1 by a trans-acting factor. PNAS USA 82:2277–2281.

Fukushima Y, Takaheshi H, et al. (1995), Extraordinary high rate of HTLVII seropositive in intravenous drug abusers in South Vietnam. AIDS Res Human Retrovirus 11:637–645.

Gaudry G, Gachou F, et al. (2002), The complementary strand of the human T-cell leukemia virus type 1 RNA genome encodes a bZIP transcription factor that downregulates viral transcript. J Virol 76:12813–12822.

Gessain A, Barin F, et al. (1985), Antibodies to human T-lymphotropic virus type-1 in patients with tropical spastic parapersis. Lancet 2: 407–410.

Ghez D, Lepellertier Y, et al. (2006), Neuropilin-1 involved in human T-cell lymphotropic virus type 1 entry. J Virol 80:6844–6854.

Gortuzzo E, Arango C, et al. (2000), Human T-cell lymphotropic virus-I in Latin America. Infect Dis Clin North Am 14: 211–239.

Helle B, Appenzeller O, et al. (1992), Chronic neurodegenerative disease associated with HTLV II infection. Lancet 339: 645–646.

Hoshino H (2012), Cellular factor involved in HTLV-1 entry and pathogenicity, Front in Microbiology; doi: 10.3389/fmicb.2012.00222.

Hottori S, Kiyokawa T, et al. (1984), Identification of gag and env gene products of human T-cell leukemia virus. Virology 136:338–347.

Ignakure T, Stirichcombe JC, et al. (2003), Spread of HTLV-1 between lymphocytes by virus-induced polarization of the cytoskeleton. Science 299:1713–1716.

Jims Q, Agrawal L, et al. (2006), Infection of CD4 T lymphocytes by the human T cell leukemia virus type 1 is mediated by the glucose transporter GLUT-1: Evidence using antibodies specific to the receptor's large extracellular domain. Virology 349:184–191.

Jones KS, Fugo K, et al. (2006), Human T-cell leukemia virus type 1 (HTLV-1) and HTLV-2 use different receptor complexes to enter T cells. J Virol 80: 8291–8302.

Jones KS, Lambert S, et al. (2012), Molecular aspects of HTLV-1 entry: Functional domains of HTLV-1 surface subunit (SU) and the relationship to the entry receptors. Viruses 3:794–810.

Jones KS, Petrow-Sadowski C, et al. (2005), Heparan sulfate proteoglycans mediate attachment and entry of human T-cell leukemia virus type 1 virions into CD4+ cells. J virol 79: 12692–12702.

Jones KS, Petrow-Sadowski C, et al. (2008), Cell-free HTLV-1 infects dendritic cells leading to transmission and transformation of CD4 (+) T cells. Nat Med 14: 429–436.

Khabbaz RF, Onorato I, et al. (1992), Seroprevalence of HTLV-1 and HTLV-2 among intravenous drug users and persons in clinics for sexually transmitted diseases. NEJM 326: 375–380.

Kim FJ, Manel N, et al. (2004), HTLV-1 and -2 envelope SU subdomains and critical determinants in receptor binding. Retrovirology 1:41.

Kusukara K, Anderson M, et al. (1999), Human T-cell leukemia type 2 Rex protein increases stability and promotes nuclear to cytoplasmic transport of gag/pol and env RNAs. J Virol 73: 8112–8119.

Lambert S, Bouttier M, et al. (2009), HTLV-1 uses HSPG and neropilin-1 for entry by molecular mimicry of VEGF165. Blood 113: 5176–5185.

Lavorgna A, Harhaj EW (2014), Regulation of HTLV-Tax stability, cellular trafficking, and NF-kB activation by the ubiquitin-proteasome. Viruses 6:3925–3943.

Lee H, Swanson P, et al. (1989), High rate of HTLV II infection in seropositive IV drug abusers from New Orleans. Science 244: 471–47.

Manel N, Kim FJ, et al. (2003), The ubiquitous glucose transporter GLUT-1 is a reporter for HTLV. Cell 115:449–459.

Mathieux R, Gessain A (2009), The human HTLV-3 and HTLV-4 retrovirus: New members of the HTLV family. Pathol Biol 57:161–166.

Matsuoka M, Jeang KT (2011), Human T-cell leukemia virus type 1 (HTLV-1) and leukemic transformation: Viral infectivity, Tax, HBX, and therapy. Oncogene 30: 1379–1389.

Mitobe Y, Yasunaga J, et al. (2015), HTLV-1 bZIP factor RNA and protein impart distinct functions on T-cell proliferation and survival. Cancer Res 75:4143–4152.

Mitagami Y, Yasunaga J-I, et al. (2015), Interferon-y promotes interaction and development of T-cell lymphoma in HTLV-1 bZIP factor transgenic Mice. PLoS Pathog 11: e1005120.

Mohr CF, Gross C, et al. (2015), Regulation of the tumor marker Fascin by the viral oncoprotein Tax of human T-cell leukemia virus type-1 (HTLV-1) depends on promoted activation and a promoter-independent mechanisms. Virology 485:481–491.

Navayan M, Kusuhara K, Green L (2001), Phosphorylation of two serine residues regulates human T-cell leukemia virus type 2 Rex function. J Virol 75:8440–8448.

Njemeddine M, Barnard AL, et al. (2005), Human T-lymphotyopic virus, type I, tax protein triggers microtubule reorientation in the virological synapse. J Biol Chem 280:296533–29660.

Nicot C, Harrod RL, et al. (2008), Human T-cell leukemia/lymphoma virus type 1 nonstructural genes and their function. Oncogene 24:6026–6034.

Overbaugh J (2004), HTLV-1 sweet-talks its way into cells. Nat Med 10:20–21.

Overbaugh J, Miller AD, Eiden MV (2001), Receptors and entry cofactors for retroviruses include single and multiple transmembrane-spanning-proteins as well as a newly described glycophosphatidylinositol-anchored and secreted particles. Microbiol Mol Biol Rev 65: 371–389.

Pinon JD, Klasse PJ, et al. (2003), Human T-cell leukemia virus type I envelope glycoprotein gp46 interacts with cell surface heparan sulfate proteoglycans. J Virol 77: 9922– 9930.

Pique C, Jones KS (2012), Pathways of cell-cell transmission of HTLV-1. Front Microbiol 3:e378.

Pique C, Pham D, et al. (1992), Human T-cell leukemia virus type 1 envelope protein maturation process: Requirements for syncytium formation. J Virol 66:906–913.

Proietti EA, Carneiro-Proietti AB, et al. (2007), Global epidemiology of HTLV-1 infection and associated disease. Oncogene 24: 6058–6068.

Satou Y, Utsunomiya A, et al. (2012), HTLV-1 modulates the frequency and phenotype of Fox P3+CD4+ T cells in virus-infected individuals. Retrovirology 9:46.

Satou Y, Yasunaga J, et al. (2006), HTLV-1 basic leucin zipper factor gene mRNA supports proliferation of adults T cell leukemia cells. PNAS USA 103:720–725.

Satou Y, Yasunaga J, et al. (2011), HTLV-1 bZIP factor induces T-cell lymphoma and systemic inflammation in vivo. PLos Pathog 7: e1001274.

Silverman LR, Phipps AJ, et al. (2004), Human T-cell lymphotropic virus type 1 open reading frame II encoded p30II is required for in vivo replication: evidence of in vivo reversion. J Virol 78: 3837–3845.

Sodroski JG, Rosen CA, Haseltine WA (1984), Trans-acting transcriptional activation of the long terminal repeat of human T lymphotropic virus in infected cells. Science 223:381–385.

Sonoda S, Li HC, Tajina K (2011), Ethnoepidemiology of HTLV-1 related disease: Ethnic determinants of HTLV-1 susceptibility and its worldwide dispersal. Cancer Sci 102: 295–301.

Sugata K, Satou Y, et al. (2012), HTLV-1 bZIP factor impairs cell-mediated immunity by suppressing product of Th1 cytokines. Blood 9:434–444.

Takeda S, Maeda M, et al. (2004), Genetic and epigenetic inactivation of tax gene in adults T-cell leukemia cells. Int J Cancer 109:559–567.

Tanaka A, Jinno-Oue A, et al. (2012), Entry of human T-cell leukemia virus type 1 augmented by heparan sulfate proteoglycans bearing short heparin-like structure. J Virol 86:2959–2969.

Taniguchi Y, Nosaka K, et al. (2008), Silencing of human T-cell leukemia virus type 1 gene transcription by epigenetic mechanism. Retrovirology 2:64.

Tordjman R, Lepelletier Y, et al. (2002), A neuronal receptor, neuropilin-1 is essential for the initiation of the primary immune responses. Nat Immunol 3:477–482.

Uretaa-Videl A, Angelin-Duclos C, et al. (1999), Mother-to-child transmission of human T-cell leukemia/lymphoma virus type I: Implication of high antiviral antibody tier and high proviral load in carrier mothers. Int J Cancer 82:832–836.

Yamamoto-Taguchi N, Satou Y, et al. (2013), HTLV-1 bZIP factor induces inflammation through labile Foxp3 expression. PLos Pathog 9: e1003630.

Ye J, Silverman L, et al. (2003), HTLV-1 Rex is required for viral spread and persistence in vivo but is dispensable for cellular immortalization in vitro. Blood 102:3963–3969.

Zhao T, Matsuoka M (2012), HBZ and its role in HTLV-1 oncogenesis. Front Microbiol doi:10.3389/fmicb.2012.00247.

CHAPTER 4

ONCOVIRAL INTEGRATION: EBV AS A MODEL

Introduction

Viral gene integration into the host's chromosome is an essential step for the successful completion of the life cycle of several viruses, such as retroviruses and adeno-associated viruses, but herpesvirus genomes are maintained as extrachromosomal circular episomes in the nucleic acid of infected cells without needing integration. However, a number of studies have reported of chromosomally integrated herpesvirus (CIHHV) DNA, which means that herpesvirus can integrate into the host's chromosome in certain circumstances.

Furthermore, a virus such as HHV-6 has been found to integrate into the germ lines of about 1 per cent of the global population, suggesting that integration may represent either a sporadic or anecdotal event. Integration causes instability in the host's genes, as shown by elevated adjacent mutation rates, with the viral genome consisting of subgenomic fragments; therefore, there is no possibility of producing infectious viral particles. This chapter will review the latest data on EBV integration, the consequences of integration, and the methods utilised in analysing integration.

4.1   EBV Genome and Site of Integration

Epstein-Barr virus (EBV) is a human herpesvirus with the ability to immortalise human B lymphocytes in vitro. In EBV-infected B cells, the virus is usually in episomal state, forming multiple copies of covalently closed circles. A number of malignant lymphoma such as Burkitt's lymphoma, nasopharyngeal carcinoma, and Hodgkin's disease are associated with EBV infection. In individuals lacking efficient T-cell function, for example, in AIDS patients or transplant recipients, EBV-immortalised B lymphocytes can grow into immunoblastic lymphoma. In its 2008 classification of hematopoietic and lymphoid-associated tumors, the WHO recognised another entity in the elderly called EBV-positive diffuse large B-cell lymphoma (DLBCL). Chronic active EBV infection (CAEBV) is associated with prolonged fever, wasting, hepatosplenomegaly, and cytopenia. In some cases, patients with CAEBV also develop a fulminate course with lymphoid malignant.

The 170kb of EBV genome (figure 4. 1) is a linear ds DNA and contains at least eighty-six ORFs. The genome contains a long unique region, which is interspersed by four major internal repeats (IR1 to IR4) and terminal repeat (TR). Nine latent proteins including Epstein-Barr nuclear antigen 1 (EBNA1), EBNA2, EBNA3A, -3B,-3C, EBNA-LP and latent membrane protein 1 (LAMP1), and LAMP2A, -2B are encoded by genes located in the unique region of the genome. Other ORFs have been reported which encode capsid protein, transcriptional factors, and lytic proteins of various functions. Furthermore, in addition to the protein-coding genes, the EBV genome also encodes other non-coding RNAs such as EBV-encoded small RNA1 (EBER1) and 2 (EBER2), BART-driven microRNAs (mRNAs-BART), and BHRF1 microRNAs (miRNAs-BHRF).

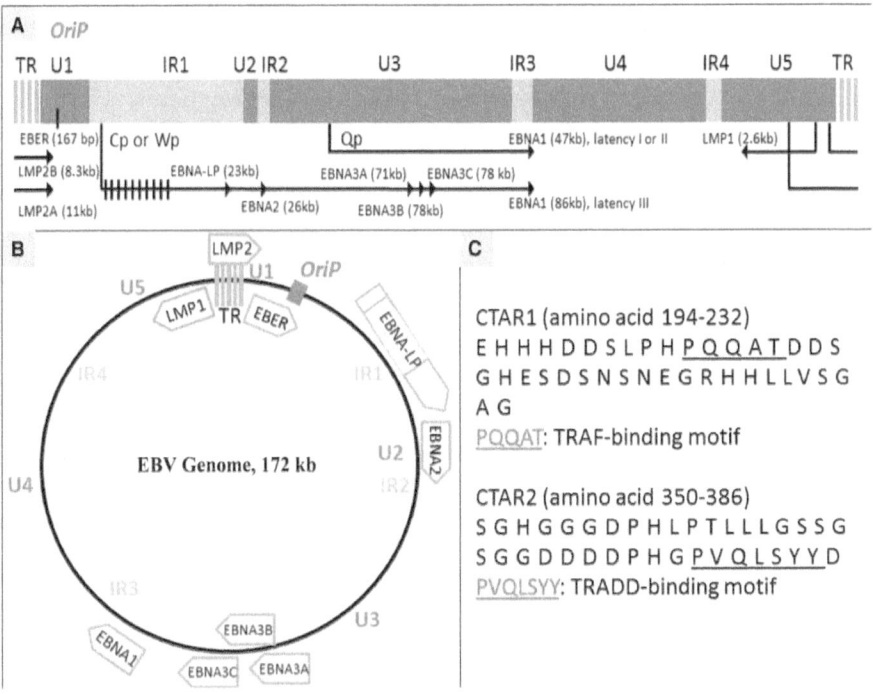

Figure 4.1: Schematic diagram of linear EBV genome

Four complete or partial EBV genome have been described as B95-8, AG876, GD1, and GD2. B95-8 was the first complete genome to be sequenced, and it was derived from an individual with infectious mononucleosis. AG876 originated from Burkitt's lymphoma in a case from Ghana. It is the only complete type 2 EBV sequence available to date. GD1 and GD2 are EBV genomes derived from NPC patients from the Guangdong Province in southern China. GD1 was isolated from saliva of NPC patients; GP2 was isolated from saliva of NPC patients, while GD2 was isolated from an NPC tumour. Cis-acting elements which mediate DNA replication during latency have been identified as Ori-P (for plasmid origin of replication). The viral DNA replication takes place once per cell cycle; it proceeds bidirectionally from the Ori-P and is dependent on cellular proteins and EBNA1, with studies showing that EBNA1 binding to Ori P is essential for plasmid DNA replication and episome maintenance. It can also function as a transcriptional enhancer of the C promoter (Cp). Another origin of replication different from Ori P has been described and referred to as Ori Lyt. It is associated with amplification of the viral genome. Replication assay showed that seven EBV proteins are required for Ori Lyt-dependent replication: DNA polymerase (BALF5), polymerase processivity factor (BMRF1), single-stranded DNA binding protein (BALF2), Primase (BSLF1), Helicase (BBLF4), Helicase/primase associated protein (BBLF2/3), and EB4. In addition, several non-essential proteins with enzymatic activities involved in biochemical pathways are also encoded by the virus.

Integration of EBV is an essential mechanism for persistence and for viral interaction with cellular genes, especially with those genes involved the regulation of cell growth and tumorigenesis. Elucidating the site of integration is essential for better understanding the mechanism of persistence in EBV-associated malignancies. Analysis of integrated EBV DNA is complicated as a result of highly methylated DNA, which hinders mapping of EBV genome, and multiple copies of viral episomes, which gives interfering noise at the EBV integrated sites.

However, a number of studies have been undertaken to elucidate EBV integration sites and the mechanism of integration in EBV infection. A study by Luo et al. using NAB-2 cell line reported that EBV was integrated via the terminal repeat, and the integration site was located in chromosome 2p13 between two oncogenes, *REL* and *BCL11A*. Others have reported of integration of EBV in chromosome 6 in Raji cell line, resulting in loss of BACH2 gene. As to whether integration occurs randomly or not, the data is still debatable, but a study by Lestou et al. reported that EBV integration is nonrandom, with the involvement of bands 1p31, 1q43, 2p22, 3q28, 4q13, 5p14, 5q12, and 11p15 in most of the cell lines. One interesting study reported that EBV integration was in G-band-positive materials. This band refers to regions in the chromosome that stain Giemsa reagents and generally associated with heterochromatin, a region associated with many repeats and no functional genes.

It must be pointed out that integration does not occur exclusively in regions without genes, as integration sites have been reported to occasionally overlap within bona fide cellular genes, such as MACF1 in Namalwa cells,

BACH2, and BCL11A. From this data, it can be concluded that EBV is integrated at different sites. Integration of the virus into the host genome leads to novel fusion transcripts or local genomic instability, resulting in secondary deletions, rearrangements, duplications, or inversion of the host and viral genomic sequences. In addition, integration is associated with tumorigenesis.

4.2   Methods of Identifying Viral Integration

Mapping of oncoviral integration sites is a powerful tool for identifying cellular oncogenes. In a study, Copeland and Jenkins used a retrovirus to identify potential oncogenes by determining the viral integration sites in tumour tissues. This led to the development of a database of cancer-associated genes. Earlier methods used in mapping integration sites utilised the concept of PCR-based capture and amplification assays but were inefficient and highly labour intensive. High-throughput generation sequencing technologies were also utilised, which led to efficient identification of integration sites. Peter et al. and other researchers developed web-based bioinformatics tools which facilitated the identification integration sites by mapping the sequence data derived from Sanger technology.

However, the tools are not sufficient to quickly map and characterise integration sites in high-throughput methods. Peter et al. introduced a new methodology which quickly maps integration sites to a reference genome from extremely large datasets. The method utilises Seqmap 2.0 and provides scalable method for sequencing, matching, clustering, and alignment, and also addresses the challenges of 454 pyrosequencing data output, namely base stutter and redundant coverage of each integration site. The Seqmap 2.0 workflow has three phases: 1. Sequencing processing, which includes identification and masking of vector features and distribution of sequence reads into multiple identifiers/barcode-specific groups. 2. Sequencing clustering and alignment. 3. Data visualisation and storage for future analysis. Figure 2 shows a typical representation of mapped integration sites in sequence viewer.

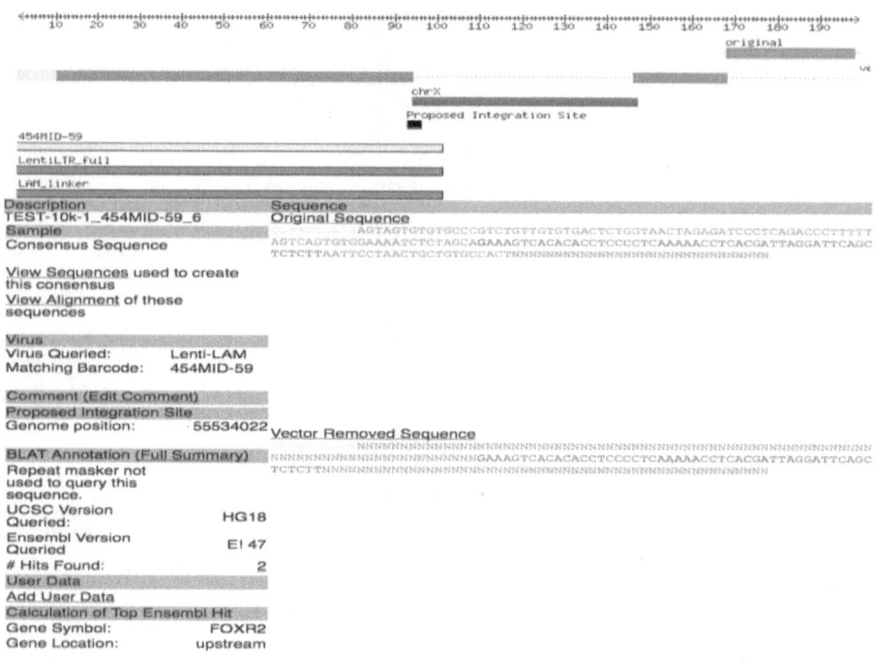

Figure 4. 2: Graphical representation of mapping
integration sites in sequence viewer

The Seqmap 2.0 can analyse data from major PCR techniques such as ligase-mediated PCR (LM-PCR) and non restrictive LAM-PCR (nrLAM-PCR). This methodology allows users to upload full sets of 454 pyrosequencing reads, create savable lists of barcodes and identifiers, create savable lists of vector features to mask from each read, and identify the appropriate reference genome to which the integration site could be mapped. Other approaches in identifying integration sites include ViralFusion Seq, Virus Finder, and Virana. Recently, Wang et al. introduced a new approach that detects virus integration sites through reference sequence customisation (VERSE).

REFERENCES

Akagi K, Li J, et al. (2014), Genome-wide analysis of HPV integration in human cancers reveals recurrent, focal genomic instability. Genome Res 24: 185–199.

Akagi K, et al. (2004), RTCGD: Retroviral tagged cancer gene database. Nucleic Acid Red 32:D523–D527.

Altmann P, Pich D, et al. (2006), Transcriptional activation by EBV nuclear antigen 1 is essential for the expression of EBV's transforming genes. PNAS USA 103:4654–4661.

Appelt JY, et al. (2009), Quickmap: A public tool for large scale gene therapy vector insertion site mapping and analysis. Gene Ther 16:885–893.

Baer R, Bankier AT, et al. (1984), DNA sequencing and expression of the B95-8 Epstein-Barr virus genome. Nature 310: 207–211.

Buchberg AM, et al. (1990), Evi-2, a common integration site involved in murine myeloid leukomagenesis. Mol Cell Biol 10:4658–4666.

Chen SJ, Chen GH, et al. (2010), Characterization of Epstein-Barr virus miRNAome in nasopharyngeal carcinoma by deep sequencing. PLoS One 5: e12745.oding a zinc-finger protein, Adv cancer Res; 54:141–157.

Copeland NG, Jenkins NA (1990), Retroviral integration in murine myeloid tumors to identify Evi-1, a novel encoding a zinc-finger protein. Adv cancer Res 54:141–157.

Gabriel R, et al. (2009), Comprehensive genomic access to vector integration in clinical gene therapy. Nat Med 15: 1431–1436.

Giordano FA, et al. (2007), New bioinformatics as strategies to rapidly characterize retroviral integration site of gene therapy vector. Methods Inf Med 40:542–547.

Hawkins TB, Dantzer J, et al. (2011), Identifying viral integration sites using Seqmap 2.0, Bioinformatics 27:720–722.

Ishihara S, Okada S, et al. (1995), Chronic active Epstein-Barr virus infection in children in Japan. Acta Pediatr 84:1271–1275.

Lestou VS, De Braekeleer M, et al. (1993), Non-random integration of Epstein-Barr virus in lymphoblastoid cell lines. Genes Chromosome Cancer 8:38–48.

Li J-W, Wan R, Yu C-S, et al. (2013), ViralFusion Seq: Accurately discover viral integration events and reconstruct fusion transcripts at single-base resolution. Bioinformatics 29:649–51.

Liu P, Tang X, et al. (2011), Direct sequencing and characterization of a clinical isolate of Epstein-Barr virus from nasopharyngeal carcinoma tissue using next generation sequencing technology. J Virology 85:11291–11299.

Lu F, Wilkramasinghe P, et al. (2010), Genome-wide analysis of host-chromosome binding sites for Epstein-Barr virus nuclear antigen 1 (EBNA1). Virology J; 7:62.

Luo W-J, Takakuwa T, et al. (2004), Epstein-Barr virus is integrated between REL and BCL-11A in America Burkitt lymphoma cell line (NAB-2). Lab Invest 84:1193–1199.

Knipe DM, Howley PM, et al. (2007), *Field's virology*. Lippincott Williams and Wilkins.

Kwok H, Tang AHY, et al. (2012), Genome sequencing and comparative analysis of Epstein-Barr virus genome isolated from primary nasopharyngeal carcinoma biopsy. PLoS One 7:e36939.

Matsuo T, Heller M, et al. (1984), Persistence of the entire Epstein-Barr virus genome integrated into human lymphocyte DNA. Science 226:1322–1325.

Parker BD, Bankier A, et al. (1990), Sequence and transcription of Raji Epstein-Barr virus DNA spanning the B95 deletion region. Virology 179:339–346.

Peter B, et al. (2008), Automated analysis of viral integration sites in gene therapy research using the Seqmap web resources. Gene Ther 15:1294–1298.

Pizzo PA, Magrath IT, et al. (1978), A new tumor-derived transforming strain of Epstein-Barr virus. Nature 272: 629–631.

Polan A, Addison C, et al. (2006), The genome of Epstein-Barr virus type 2 strain AG876. Virology 350:164–170.

Portes-Sentis S, Sergeant A, Gruffat H (1997), A particular DNA structure is required for the function of a cos-acting component of the Epstein-Barr virus OriLyt origin of replication. Nucleic Acid Research 7:1347–1354.

Rickinson AB (1986), Chronic, symptomatic Epstein-Barr infection. Immunology Today 7:13–14.

Schelhorn S-E, Fischer M, et al. (2013), Sensitive detection of viral transcripts in human tumor transcriptomes. PLoS Comput Biol 9:e1003228.

Schmidt M, et al. (2003), Efficient characterization of retro-, lenti-, and foamy vector-transduced cell population by high-accuracy insertion site sequencing. Ann NY Acad Sci 996: 112–121.

Schmidt M, et al. (2007), High resolution insertion site analysis by linear amplification-mediated PCR (LAM-PCR). Nat Methods 4:1051–1057.

Smith DR (1992), Ligation-mediated PCR of restriction fragment for large DNA molecules. PCR Methods Appl 2: 21–27

Sung W-K, Zheng H, et al. (2012), Genome-wide survey of recurrent HBV integration in hepatocellular carcinoma. Nat Genet 44:765–769.

Swaminathan S (2008), Noncoding RNAs produced by oncogenic human herpesvirus. J Cell Physiol 216: 321–326

Takakuwa T, Luo W-J, et al. (2004), Integration of Epstein-Barr virus into chromosome 6q15 of Burkitt's lymphoma cell line (Raji) induces loss of BACH2 expression. Am J Pathol 164:967–974.

Tarbouriech N, Buisson M, et al. (2006), Structural genome of the Epstein-Barr virus. Acta Crystallogr D Biol Crystallogr 62: 1276–1285.

Wang Q, Jia P, Zhao Z (2013), Virus Finders: Software for efficient and accurate detection of viruses and their integration sites into host genome through next generation sequencing data. PLos One 8:e64465.

Weiss LM, Movahed LA, et al. (1989), Detection of Epstein-Barr viral genomes in Reed-Stemberg cells of Hodgkin's disease. NEJM 32:502–506.

Yates JL, Warren N, et al. (1984), A Cis-acting element for Epstein-Barr viral genome that permits stable replication of recombinant plasmids in latently infected cells. PNAS USA 81: 3806–3810.

Zeng MS, Li DJ, et al. (2005), Genomic sequence analysis of Epstein-Barr virus strain GD1 from a nasopharyngeal carcinoma patient. J Virology 79:15323–15330.

Zimber U, Adldinger HK, et al. (1986), Geographical prevalence of two types of Epstein-Barr virus. Virology 154:56–66.

CHAPTER 5

PATHOGENESIS OF HIV-ASSOCIATED CANCER

Introduction

Lymphomas are part of the most critical complications associated with HIV infection; they occur in high frequencies and are significant cause of morbidity and mortality. Most are aggressive B-cell lymphomas, which are histologically heterogeneous. They include lymphoma commonly diagnosed in HIV-negative patients and others that are primarily associated with HIV infection and common in patients with severe immunodeficiency. These include diffuse large B-cell lymphoma (DLBCL), which includes primary CNS lymphoma (PCNSL), and Burkitt's lymphoma (BL). Primary effusion lymphoma (PEL), plasmablastic lymphoma, and classic Hodgkin lymphoma also manifest in HIV infection but are less frequent. Other lymphomas such as follicular lymphoma and peripheral T-cell lymphoma have also been reported but are not common. Epidemiologic data suggest that HIV-positive patients are 60- to 200-fold at risk of developing non-Hodgkin lymphoma (NHL), the majority of which are DLBLC. The introduction of combination antiretroviral therapy (CART) saw a reduction in HIV-associated lymphomas, improved quality of life, and improved immune function; however, the risk for these cancers are still high. This chapter will review the current data on HIV-associated NHL.

6.1 Epidemiology

NHL is mostly a high-grade B-cell lymphoma. In patients with AIDS, NHL represents the second most common lymphomas after KS. Patients with HIV infection are at increased risk of developing NHL while patients, while HIV infection increases the risk of developing NHL 100 to 200 times higher than in the general population. NHL is characterised by their rapid progression, frequent extranodal initial manifestation, and poor outcome. The characteristic extranodal manifestation as clinical presentation of the disease is common in all types of AIDS-associated lymphomas. In the years before 1996, it was estimated that 3.0 to 3.6 per cent of all AIDS-defining diseases were attributed to NHL. The development of NHL is associated with a relative risk of death up to 20-fold. This means HIV-associated NHL caused up to 16 per cent of all deaths attributed to AIDS. The introduction of HAART led to reduction in AIDS-related morbidity and mortality, including NHL. A study by the AIDS Clinical Group found that incidence of both KS and NHL decreased, but the former was more profound and consistent. However, the Multicentre AIDS Cohort Study (MACS) reported that there was an increase in the incidence of NHL, while the incidence of KS fell by 66 per cent in the same year. A study within EuroSIDA, a multicentre observational cohort of more than 8,500 patients from across Europe, found that the incidence of NHL among HIV-infected patients decreased significantly after the introduction of HAART; the decline was more pronounced for primary brain lymphoma. However, after starting HAART, patients with insufficient immunologic and virologic responses were at highest risk of NHL. A study in Botswana reported that over a five-year period, incidence of KS decreased but incidence of NHL and HPV-associated cancer increased during the period. Although the incidence of NHL maybe reducing, it remains a fact that among HIV-infected individuals, the incidence of NHL is 70 times higher than in the general population. NHL is seen so frequently in HIV-infected individuals that the presence of NHL is an AIDS-defining criteria.

6.2 Pathogenesis

HIV-associated NHL is characterised by the presence of recurrent genetic alteration, which may be as a result of errors in the normal processes that occur in activated B-cells that involve modification of somatic DNA, such as immunoglobulin (Ig) class-switch recombination (CSR) and somatic hypermutation (SHM). In BL, it has been reported that c-MYC gene found on chromosome 8 is associated with translocation and then placed closely to Ig heavy chains (IgH) locus on chromosome 14 or to Ig high chain on chromosome 2 or 22, which results in the upregulation of c-MYC expression. These translocations may be due to errors in IgH CSR or SHM; they are believed to play a role in the development of BL. Similarly, DLBCL subtype of HIV-associated NHL is associated with BCL6 oncogene translocation and mutation, which might be due to error in both IgH CSR and SHM.

In a study, Deffenbacher et al. reported that HIV-associated NHL is characterised by recurrent multiple changes in the chromosome involving MYC, BCL6, and other potential pathways such as FAS, MTR, RAS, and p53. Therefore, alterations or changes in HIV infection that leads to chronic B-cell hyperactivation may contribute to the development of HIV-associated NHL. A DNA-editing enzyme, Activation-induced cytidine deaminase (AID), which is normally active during B-cell activation and essential for Ig SHM and CSR, promoted c-MYC/IgH translation. This is required for the development of lymphomas of the germinal centre (GC). However, AID also produces DNA double-strand breaks in both Ig genes and other loci, thereby causing widespread genome instability, which could contribute to a certain degree of non-Ig-related modifications seen in HIV-associated NHL.

AID expression is notably increased in PBMC of HIV-positive individuals before the diagnosis of NHL. Data shows that an increased risk of HIV-associated NHL is due to a combination of immunodeficiency, increased immune activation, and possible HIV insertional mutagenesis. Cytokines have been implicated in the pathogenesis of HIV-associated NHL. B-cell activation is an important phenomenon in the development of

HIV-associated NHL. The activation and proliferation of B lymphocytes in particular increase the possibility of chromosomal translocation and of oncogenic mutation in the process of lymphomagenesis. In HIV-associated NHL, B-cell activation represents a novel marker of future development of the lymphoma and provides an insight into the aetiology of these cancers.

One of the hallmarks of HIV infection is the elevation of Ig in HIV-infected individuals; however, a study by Landgren et al. showed that the levels of Ig in HIV-positive people who will develop NHL and those who do not are not different. But the study showed that k and λ light chains (FLCs) are elevated five years before the diagnosis of HIV-associated NHL. This means FLCs are markers of NHL development in the future. FLCs are also elevated in autoimmune disorders, such as rheumatoid arthritis and Sjögren syndrome, as well as other non-HIV-related lymphoproliferative disorders. This confirms their importance as markers of future or present B-cell proliferation.

Years ago, a number of markers were associated with B-cell lymphomagenesis in HIV infection. These include soluble CD27 (sCD27) and sCD30, who are members of the tumour necrosis factor receptor (TNF-R) superfamily. Both are markers of B-cell and T-cell activation. They are also expressed in HIV-negative NHL. A study by Widney et al. reported that HIV-positive subjects have decreased CD27 expression on circulating B-cells; this was inversely correlated with serum sCD27 levels. sCD30 levels were also associated with other activation markers, such as IL10, IL6, sCD23, sCD27, and CXCL3, with high sCD30 levels associated with poor survival in HIV-associated NHL. Several studies have shown that there is an association between the cytokine genetic signature and the development of HIV-associated NHL.

For example, production of high levels of IL10 was associated with increased risk of developing HIV-associated NHL, while TNF-α was found to elevated in HIV-associated NHL post-diagnosis. Similarly, macrophages have been implicated in the pathogenesis of HIV-associated NHL. Macrophages form a large portion of inflammatory infiltrate in most, if not all, cancers. It has been reported that tumour-associated macrophage (TAM) constitute about 80 per cent of the total tumour mass, depending on the type of tumour. Evidence supports the argument that macrophage actively promotes tumour progression in numerous human cancers. TAM plays an essential role in every hallmark step of tumour progression, with a proposal put forward that macrophages might play a direct role in cancer progression, as some studies reported that metastatic cells arise from cells of myeloid/macrophage lineage. In HIV-associated NHL, it has been reported that TAM have a negative impact on NHL tumour grade and patient survival.

Follicular lymphoma(FL) is the second most common type of NHL in HIV-negative population. It is a clinically heterogeneous disease, with survival ranging from two to twenty years after diagnosis. In a study of FL patients, it was found that those with high TAM levels had a median OS of only five years as compared to those with low TAM, who had an OS of 16.3 years. Another study showed that in other forms of B cell HLS, the TAM content increases with malignancy grade and is highly correlated with tumour vascularity. However, other studies suggested that TAM did not correlate with progression-free survival or tumour grade in patients with DLBCL. Although further studies are required to elucidate the role of TAM in DLBCL pathogenesis, data available suggest that TAM in HIV-negative B cell NHLs have a negative impact on progression, as found in other cancer types. Therefore, it will be of interest to determine if TAM levels are elevated in HIV-positive NHL in comparison to HIV-negative NHL.

REFERENCES

Aissani B, et al. (2009), The major histocompatibility complex conserved extended haplotype 8.1 in AIDS-related non-Hodgkin lymphoma. J Acquir Immune Defic Syndr 52:170–179.

Bellerini P, et al. (1992), Molecular pathogenesis of HIV-associated lymphomas. AIDS Res Hum Retroviruses 8:731–735.

Beral V, et al. (1991), AIDS-associated non-Hodgkin lymphoma. Lancet 337:305–309.

Bingle L, et al. (2002), The role of tumor-associated macrophages in tumor progression: Implications for new anticancer therapies. J Pathol 196:254–265.

Bornkamm GW (2009), Epstein-Barr virus and the pathogenesis of Burkitt's lymphoma: More questions than answers. Int J Cancer 124:1745–1755.

Bottazzi B, et al. (1983), Regulation of the macrophage content of neoplasms by chemoattractants. Science 220:210–212.

Breen EC, et al. (1999), The development of AIDS-associated Burkitt's/small noncleaved cell lymphoma is preceded by elevated serum levels of interleukin. Clin Immunol 92:293–299.

Breen EC, et al. (2005), Elevated levels of soluble CD44 precede the development of AIDS-associated non-Hodgkin's B-cell lymphoma. AIDS 19:1711–1712.

Breen EC, et al. (2003), Non-Hodgkin's B cell lymphoma in persons with acquired immunodeficiency syndrome is associated with increased serum levels of IL10, or the IL10 promoter-592 C/C genotype. Clin Immunol 109:119–129.

Buchbinder S, Holmberg S, et al. (1999), Combination antiretroviral therapy and incidence of AIDS-related malignancy. J AIDS 21 (Suppl 1): S23–26.

Burdin N,et al. (1995), Endogenous IL-6 and IL-10 contribute to the differentiation of CD40-activated human B lymphocytes. J Immunol 154:2533–2544.

Cagigi A, et al. (2008), Altered expression of the receptor-ligand pair CXCR5/ CXCL13 in B cells during chronic HIV-1 infection. Blood 112:4401–4410.

Carbone A, et al. (2009), HIV-associated lymphomas and gamma-herpesviruses. Blood 113:1213–1224.

Chiarle R, et al. (1990), CD30 in normal and neoplastic cells. Clin Immunol 90:157–164.

Cortis M, et al. (2011), Non-Hodgkin lymphoma of the oral cavity in AIDS patients in a reference hospital of infectious diseases in Argentina: Report of eleven Cases and review of the literature. J Gastrointestinal Cancer 42: 143–148.

Cramer T, Yamanishi Y, et al. (2003), HIF-1α is essential for myeloid cell-mediated inflammation. Cell 112:645–657.

Croft M (2003), Co-stimulatory members of the TNFR family: Keys to effective T-cell immunity? Nat Rev Immunol 3:609–620.

Deffenbacher KE, et al. (2010), Recurrent chromosomal alterations in molecularly classified AIDS-related lymphomas: An integrated analysis of

DNA copy number and gene expression. J Acquir Immune Defic Syndr 54:18–26.

De Milito A, et al. (2000), Loss of memory (CD27) B lymphocytes in HIV-1 infection. AIDS 15:957–964.

Dryden-Peterson S, et al. (2015), Cancer incidence following expansion of HIV treatment in Botswana. PLoS One 10:e0135602.

Ekstrom Smedby K, et al. (2008), Autoimmune disorders and risk of non-Hodgkin lymphoma subtypes: A pooled analysis within the InterLymph Consortium. Blood 111:4029–4038.

Elgert K D, et al. (1998), Tumor-induced immune dysfunction: the macrophage connection. J Leukoc Biol 64:275–290.

Engels EA, et al. (2003), Cancer risk in people infected with human immunodeficiency virus in the United States. Int J Cancer 123:187–194.

Epeldegui M, et al. (2007), Elevated expression of activation induced cytidine deaminase in peripheral blood mononuclear cells precedes AIDS-NHL diagnosis. AIDS 21:2265–2270.

Epeldegui M, et al. (2013), HIV-associated immune dysfunction and viral infection: Role in the pathogenesis of AIDS-related lymphoma. Immunol Res 48:72–83.

Duff DK, et al. (2000), The cytokine milieu of HIV-associated non-Hodgkin's lymphoma favors aggressive tumours. AIDS 14:92–94.

Dunleavy K, Wilson WH (2012), How I treat HIV-associated lymphoma. Blood 119: 3245–3255.

Fiorentino DF, et al. (1989), Two types of mouse T helper cell. IV. Th2 clones secrete a factor that inhibits cytokine production by Th1 clones. J Exp Med 170:2081–2095.

Fischer L, et al. (2009), CXCL13 and CXCL12 in central nervous system lymphoma patients. Clin Cancer Res 15:5968–5973.

Gordon J, et al. (1989), Regulation of resting and cycling human B lymphocytes via surface IgM and the accessory molecules interleukin-4, CD23 and CD40. Immunology 68:526–531.

Gordon J, et al. (1989), CD23: A multi-functional receptor/lymphokine? Immunol Today 10:153–157.

Goswami S, Sahai E, et al. (2005), Macrophages promote the invasion of breast carcinoma cells via a colony-stimulating factor-1/epidermal growth factor paracrine loop. Cancer Res 65:5278–5283

Gottenberg JE, et al. (2007), Serum immunoglobulin free light chain assessment in rheumatoid arthritis and primary Sjogren's syndrome. Ann Rheum Dis 66:23–27.

Grulich AE, et al. (2007), Incidence of cancers in people with HIV/AIDS compared with immunosuppressed transplant recipients: A meta-analysis. Lancet 370:59–67.

Herbelin A, et al. (1994), Soluble CD23 potentiates interleukin-1-induced secretion of interleukin-6 and interleukin-1 receptor antagonist by human monocytes. Eur J Immunol 24:1869–1873.

Huysentruyt L C, et al. (2008), Metastatic cancer cells with macrophage properties: Evidence from a new murine tumor model. Int J Cancer 123:73–84.

Huysentruyt LC, McGrath MS (2010), The role of macrophages in the development and progression of AIDS-related non-Hodgkin lymphoma. J Leukoc Biol 87:627–632.

Kim CH, et al. (2004), Unique gene expression program of human germinal center T helper cells. Blood 104:1952–1960.

Jacobsen L, et al. (1999), Impact of potent antiretroviral therapy on the incidence of Kaposi's sarcoma and non-Hodgkin's lymphoma among HIV-1 infected individuals. J AIDS 21 (Suppl 1): S34–41.

Ji Y, Zhang W (2010), Th17 cells: positive or negative role in tumor? Cancer Immunol Immunother 59:979–987.

Patke CL, Shearer WT (2000), gp120- and TNF-alpha-induced modulation of human B cell function: Proliferation, cyclic AMP generation, Ig production, and B-cell receptor expression. J Allergy Clin Immunol 105:975–982.

Klapproth K, Wirth T (2010), Advances in the understanding of MYC-induced lymphomagenesis. Br J Haematol 149:484–497.

Klein G, Klein E, Kashuba E (2010), Interaction of Epstein-Barr virus (EBV) with human B-lymphocytes. Biochem Biophys Res Commun 396:67–73.

Kirk O, et al. (2001), Non-Hodgkin lymphoma in HI-infected patients in the era of highly active antiretroviral therapy. Blood 98:3406–3412.

Kryczek I, et al. (2007), Cutting edge: Th17 and regulatory T cell dynamics and the regulation by IL-2 in the tumor microenvironment. J Immunol 178:6730–6733.

Landgren O, et al. (2010), Circulating serum free light chains as predictive markers of AIDS-related lymphoma. J Clin Oncol 28:773–779.

Ledergerber B, et al. (1999), Risk of HIV related Kaposi's sarcoma and non-Hodgkin's lymphoma with potent antiretroviral therapy: Prospective cohort study. BMJ 319:23–24.

Lewis C E, Pollard J W (2006), Distinct role of macrophages in different tumor microenvironments. Cancer Res 66:605–612.

Mantovani A, Sozzani S, et al. (2002), Macrophage polarization: Tumor-associated macrophages as a paradigm for polarized M2 mononuclear phagocytes. Trends Immunol 23:549–555.

Martinez-Maza O, et al. (1987), Infection with the human immunodeficiency virus (HIV) is associated with an in vivo increase in B lymphocyte activation and immaturity. J Immunol 138:3720–3724.

Martin W, et al. (2007), Serum-free light chain: A new biomarker for patients with B-cell non-Hodgkin lymphoma and chronic lymphocytic leukemia. Transl Res 149:231–235.

Mocroft A, Vella S, et al. (1998), Changing patterns of mortality across Europe in patients infected with HIV-1. EuroSIDA Study Group. Lancet 352:1725–1730.

Munzarova M, et al. (1992), Are advanced malignant melanoma cells hybrids between melanocytes and macrophages? Melanoma Res 2:127–129.

Murdoch C, Giannoudis A, Lewis C E (2004), Mechanisms regulating the recruitment of macrophages into hypoxic areas of tumors and other ischemic tissues. Blood 104:2224–2234.

Navarro JT, et al. (2000), Increased serum levels of CD44s and CD44v6 in patients with AIDS-related non-Hodgkin's lymphoma. AIDS 14:1460–1461.

O'Sullivan C, Lewis C E, et al. (1993), Secretion of epidermal growth factor by macrophages associated with breast carcinoma. Lancet 342:148–149.

Palella F, Delaney K, Moorman A, et al. (1998), Declining morbidity and mortality among patients with advanced human immunodeficiency virus infection. NEJM 338:853–860.

Pasqualucci L, et al. (2008), AID is required for germinal center-derived lymphoma-genesis. Nat Genet 40:108–112.

Patel P, et al. (2008), Incidence of cancers among HIV-infected persons compared with the general population in the United States, 1992–2003. Am Intern Med 148:728–736.

Peters BS, et al. (1991), Changing disease patterns in patients with AIDS in a referral centre in the United Kingdom: The changing face of AIDS. BMJ 302:203–7.

Petruckevitch A, et al. (1998), Disease progression and survival following specific AIDS-defining conditions: a retrospective cohort study of 2048 HIV-infected persons in London. AIDS 12:1007–1013.E

Pluda JM, et al. (1993), Parameters affecting the development of non-Hodgkin's lymphoma in patients with severe human immunodeficiency virus infection receiving antiretroviral therapy. J Clin Oncol 11:1099–1107.

Polo JM, et al. (2007), Transcriptional signature with differential expression of BCL6 target genes accurately identifies BCL6-dependent diffuse large B cell lymphomas. PNAS USA 104:3207–3212.

Rabkin C, Testa M, von Roenn J (1999), Kaposi's sarcoma and non-Hodgkin's lymphoma incidence and trends in AIDS Clinical Trial Group study participants. J AIDS 21 (Suppl 1):S31–3.

Robbiani DF, et al. (2009), AID produces DNA double-strand breaks in non-Ig genes and mature B cell lymphomas with reciprocal chromosome translocations. Mol Cell 36:631–641.

Romagnani S, et al. (2009), Properties and origin of human Th17 cells. Mol Immunol 47:3–7.

Sasson SC, et al. (2010), IL-7 receptor is expressed on adult pre-B-cell acute lymphoblastic leukemia and other B-cell derived neoplasms and correlates with expression of proliferation and survival markers. Cytokine 50:58–68.

Schroeder JR, et al. (1999), Serum soluble CD23 level correlates with subsequent development of AIDS-related non-Hodgkin's lymphoma. Cancer Epidemiol Biomarkers Prev 8:979–984.

Schroeder JR, et al. (1999), Serum sCD23 level in patients with AIDS-related non-Hodgkin's lymphoma is associated with absence of Epstein-Barr virus in tumor tissue. Clin Immunol 93:239–244.

Shels MS, et al. (2011), Proportions of Kaposi-Sarcoma, selected non-Hodgkin lymphoma, and cervical cancer in the United States occurring in persons with AIDS, 1980–2007. JAMA 305:1450–1459.

Steinman L (2007), A brief history of T (H) 17, the first major revision in the T (H) 1/T (H) 2 hypothesis of T cell-mediated tissue damage. Nat Med 13:139–145.

Takagi R, et al. (2008), B cell chemoattractant CXCL13 is preferentially expressed by human Th17 cell clones. J Immunol 181:186–189.

Terrier B, et al. (2009), Serum-free light chain assessment in hepatitis C virus-related lymphoproliferative disorders. Ann Rheum Dis 68:89–93.]

Tirelli U, et al. (2000), Epidemiological, biological and clinical feature of HIV-related lymphomas in the era of highly active antiretroviral therapy. AIDS 14: 1675–1688.

van Baarle D, et al. (2001), Dysfunctional Epstein-Barr virus (EBV)-specific CD8(+) T lymphocytes and increased EBV load in HIV-1 infected individuals progressing to AIDS-related non-Hodgkin lymphoma. Blood 98:146–155.

van Oers MH, et al. (1993), Expression and release of CD27 in human B-cell malignancies. Blood 82:3430–3436

Vissers JL, et al. (2001), BLC (CXCL13) is expressed by different dendritic cell subsets in vitro and in vivo. Eur J Immunol 31:1544–1549.

Widney DP, et al. (2005), Serum levels of the homeostatic B cell chemokine, CXCL13, are elevated during HIV infection. J Interferon Cytokine Res 25:702–706.

Widney DP, et al. (2010), Expression of the B cell chemokine, CXCL13, in AIDS-associated non-Hodgkin's lymphoma, AIDS Res Treat; 2010: 164586 (In press).

Widney D, et al. (1999), Aberrant expression of CD27 and soluble CD27 (sCD27) in HIV infection and in AIDS-associated lymphoma. Clin Immunol 93:114–123.

Wolf T, et al. (2005), Changing incidence and prognostic factors of survival in AIDS-related non-Hodgkin lymphoma in the era of highly active antiretroviral therapy (HAART). Leuk Lymphoma 46:207–215.

Wong HL, et al. (2010), Cytokine signaling pathway polymorphisms and AIDS-related non-Hodgkin lymphoma risk in the multicenter AIDS cohort study. AIDS 24:1025–1033.

Yawetz S, et al. (1995), Elevated serum levels of soluble CD23 (sCD23) precede the appearance of acquired immunodeficiency syndrome-associated non-Hodgkin's lymphoma. Blood 85:1843–1849.

CHAPTER 6

THE CELLULAR TUMOR
ANTIGEN P53

Introduction

The p53 gene was initially considered to be an oncogenic protein. However, a study disproved this belief when it was shown that in Friend virus-induced mouse erythroleukaemia, p53 gene was a target for viral integration, resulting in its inactivation. Subsequent studies showed that it was a tumour-suppressor gene and recognised as one of the frequently inactivated genes in more than 50 per cent of all human cancers. The human p53 gene is found in chromosome 17p13, and Sothern blot analysis has revealed that there is a single p53 gene in the human genome spanning about 20 kb of the genomic DNA. The gene (figure 1) is made of eleven exons; the first exon is a non-coding exon, followed by a large first intro of 10kb in length. There are some highly conserved domains within the exons 5, 7, and 8. The mRNA transcript of p53 is about 2.8kb in length and can be found in most human cells, with the exception of cell in the GO phase.

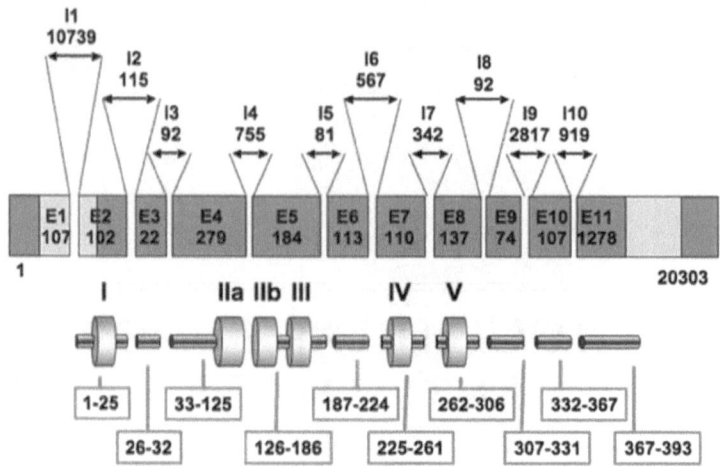

Figure 6.1: Organisation of human p53 gene

The gene encodes about 393 amino acids, with a molecular weight of 53Da. Based on its structure and function, the gene is divided into three distinct domains: 1. The transcriptional activation domain at the amino or N-terminus. 2. The central sequence specific DNA binding domain. 3. The multifunctional basic carboxyl or C-terminus. The p53 pathway is made of hundreds of genes and their products that respond to a number of intrinsic and extrinsic stress signals such as DNA damage, oncogene activation, NO production, and hypoxia. These stress signals all impact under the cellular haemostatic mechanism that monitor and control the fidelity of DNA replication, chromosome segregation, and cell division. Among the stresses that activate the p53 protein is damage to the integrity of DNA in a cell.

The p53 gene has a number of functions. These include cell cycle checkpoint, promoting apoptosis through transcription-dependent and independent mechanisms. Transcription-independent pro-apoptotic activities modulate the functions of protein involved in the apoptotic machinery, anti-apoptotic capabilities, and regulation of the senescence response, maintaining genetic stability and neurogenetic ability; it also plays a role in tumorigenesis.

6.1   Mutational Analysis of p53

The p53 gene is one of the most often mutated in human cancers. It involves mainly point mutation, resulting in amino acid substitution in the central domain of the protein, which impairs normal functions. Analysis of the mutational event that targets the gene has shown that both exogenous and endogenous mutational mechanisms are involved. Classes of DNA damage include deletion, insertion, and base substitution (either transition or transversion). Transition predominates in colon, brain, and lymphoid cancers, whereas G: C to T: A transversion are the most common substitution seen in cancer of the lungs and liver. Mutation at A: T base pair is seen more commonly in oesophageal carcinoma than in other solid tumours. Most transition in colorectal, brain tumours, leukaemia, and lymphomas are at CpG dinucleotide mutational hot spots. G to T transversion in lung, breast, and oesophageal carcinoma are dispersed among numerous codons. In liver tumours in geographic areas in which both aflatoxin (AFB) B1 and HBV are cancer risk factors, most mutations are at one nucleotide pair of codon 246. A number of investigations have indicated the presence of chronic HBV infection and p53 mutation in the same tumours. However, HBV infection alone does not influence the rate of p53 mutation.

Aflatoxin has the most influence on the prevalence of mutation. A number of studies have provided evidence (albeit inconclusive) to support the argument that a combination of aflatoxin and HBV infection may be responsible for mutation. In a prospective cohort study of 18,244 people, AFB was seen as having an etiological role in HCC, indicating a synergy between HBV and AFB. The study showed a statistically significant association in the presence of AFB and its metabolites in urine specimen, serum HBV surface antigen positivity, and HCC risk. In addition, the presence of promutagenic AFB-N[7] guanine adduction in the urine provided further evidence that AFB was activated to its electrophilic carcinogenic metabolite, AFB 8,9-oxide. Another study showed high frequency of AGG→ AGT transversion on the nontranscribed strand at p53 codon 249 in HCCs in some parts of China and Mozambique, with high incidence of HCC. This was associated possibly with high mutability of the third base codon 249 by AFB or selective growth advantage of hepatocyte clones carrying this surface 249ᵉˣ mutant in liver of chronically infected HBV individuals.

A study by Aguiler et al. to test the preferential mutability concept showed that in a human liver cell exposed to AFB, the third base in codon 249 is preferentially (but not exclusively) mutated in comparison to the immediate adjacent codons. This suggests that both preferential mutability and clone selection are involved in the development of HCC. In HPV-associated cervical cancer, HPV type 16 and 18 carcinomas have been linked with the p53 pathway. HPV type 16 and 18 both produce E6 protein, which complexes with the cellular protein AP6 and degrades p53 in the cytoplasm. This is a typical example of epigenetic p53 inactivation.

Mutation analysis supports this concept if p53 mutant were found more in HPV-negative than in HPV-positive malignancies, and IHC would demonstrate no p53 staining n HPV-positive cases and nuclear staining in HPV-negative cases with missense mutations. Reports show that mutation frequency was 100 per cent in HPV-negative and 0 per cent in HPV-positive tumour cell lines, but these results have not been validated in further studies on clinical tumours. Other studies confirmed the rarity of mutation in HPV-positive tumours and lines; however, mutation was reported in seven of thirty-nine HPV-negative carcinomas.

These studies provide evidence that dysfunction of the p53 protein might have a role in cervical carcinogenesis, but unless there are other alterations in the p53 pathway, it cannot be considered essential. In light of these data, more studies are needed to address the issue of low rate of mutation in HPV-negative tumours by searching for other p53 pathway abnormalities. Other data provides evidence of the involvement of additional factors in p53-associated HCC. A study by Chen et al. showed that mutation in p53 gene is linked to recurrence of HCC and that ZBP-89 may play a role in the nuclear accumulation of p53 protein in a subset of recurrent HCC. The ZBP-89 is a 4-zinc finger transcription factor that regulates the expression of several genes related to cell growth through binding to GC-rich DNA elements. ZBP- 89 has the ability to stabilise the p53 gene. The study suggested that co-localisation of p53 protein with ZBP-89 may define a subset of recurrent HCC that is more sensitive to treatment. Regarding the pattern of mutation, the study by Chen et al. found that mutation in exon 7 of the p53 gene accounted for 43.8 per cent of all alterations found, thereby suggesting that exon 7 should be considered a prevalent site of p53 mutation.

6.2 Anti-p53 Drugs

Given the central role played by p53 in cancer prevention and suppression as well as chemo and radio sensitisation, abrogation of p53 is essential during carcinogenesis for most cancers to occur. Therefore, directly targeting mutated p53 would be an ideal therapeutic strategy in cancer care. This intervention would potentially reduce the deleterious side effects resulting from most of the current treatment, which is based on DNA damage. Another additional advantage of such intervention is mutated p53 is frequently overexpressed and post-translationally modified in tumour cells. The cellular environment of tumour cells also favours functional p53-induced apoptosis. Therefore, any molecule or peptide chaperones that can stabilise mutated p53 weight conformation will activate the apoptosis pathway in tumour cells. Also, most chemotherapeutic agents and radiotherapy require a functional p53 pathway. Therefore, p53 chemical activators can increase the sensitivity of chemotherapy or radiotherapy. Below, a review is made of the current strategies used for pharmacological modulation of p53 protein, gene therapy, and utilisation of this gene in cancer detection and monitoring.

6.2.1 Gene Therapy Using TP53

In 1996, the first gene therapy using p53 was reported. It involved a retroviral vector comprising the wild-type p53 gene, which was under the control of actin promoter. This was injected directly into tumours of nonsmall lung cancer patients. With the development of a replication defective recombination p53 gene (Ad5CMV-p53), a number of clinical trials have been performed, with a few reaching phase III, but no approval has been granted yet.

6.2.2   Use of PRIMA1 and PhiKa083

One of the concepts is using a class of small molecules that can reactivate the wild-type functions of the mutated p53. The best studied is PRIMA-1, which entered the second phase of clinical trials in 2011. PRIMA-1 is converted to PRIMA-1^{MET}, the methylated form of the drug which works in a number of ways to restore the function of p53. One of the mechanisms is that PRIMA-1^{MET} covalently binds to and modifies the thiol groups in the central domain of the mutated gene. It causes the reactivation of the p53 gene, which enables it to regain its ability to induce apoptosis. Another Phikan083, a carboxyl derived from in silico screening of the crystal structure of pp53. By binding to mutated p53, it raises the melting temperature of the mutated p52, thereby resulting in the reactivation of its function.

6.2.3   P53 Stabiliser

Stabilising p53 is a novel strategy that can be utilised for therapeutic intervention. MDM2 is an E3 ubiquitin ligase with the ability of controlling the degradation of p53. Most tumours overexpress MDM2. It has been reported that even tumours without p53 mutation overexpress MDM2. In a study, it was shown that the nutlin which are cis-imidazoline compounds and act as antagonists of MDM2-p53 interaction binds in the pocket of MDM2 to prevent the p53-MDM2 interaction. Nutlin can activate the p53 pathway, thereby inducing cancer cells and xenograft tumours in mice to undergo cell cycle arrest, apoptosis, and growth inhibition. M1-219 is a small molecule that can inhibit the p53 pathway in cells with wild-type p53. It was shown that apoptosis and cell cycle arrest were observed in xenograft tumours, thereby resulting in tumour regression. But utilising MDM2 inhibition and p53 in normal tissue can be harmful, as reported by Ringshausen et al. in a study showing that p53 is spontaneously activated in many tissues in MDM2-deficient mice.

In addition, p53 can induce a number of pathologies. Other classes of p53 stabilisers have been described. These include RITA (reactivation of p53 and induction of tumour cell apoptosis), which was discovered by Issaeva et al. It acts by binding to p53 and inhibits the p53-MDM2 interaction, both in vivo and in vitro. RITA therefore induces apoptosis in various cancer cells that retain wild type p53. Tenovin also stabilises p53 by activating the gene. Two compounds have been identified: Tenovin-1 and Tenovin-6. Tenovin-6 is more water soluble than Tenovin-1. It has been established that Tenovin-1 inhibits protein-deacetylating activities of SirT1 and SirT2 of the Sirtuin family. Deacetylation of p53 not only stabilises p53 but also interferes with MDM2-mediated p53 degradation. This means these compounds can target p53 gene modulation and also activate or increase the activity of p53.

6.2.4   P53-Based Immunotherapy

Immunotherapy can be used as an adjuvant therapy. Although p53 frequently mutates, the remainder of the molecule keeps its wild-type sequence. Nonmutated peptides can be processed from the altered p53 molecule and presented by tumour cells for T cell recognition. CTLs are the most important effectors for antitumour immune response. A study found that adoptive immunotherapy of tumour-bearing mice with p53-specific CTL resulted in eradication of p53-overexpressed tumours in the absence of immunopathological damage to normal tissue.

These tumours also eliminated tumours that did not show greatly enhanced expression of p53. This indicates that the sensitivity of these tumours for p53-specific CTL is determined by the efficacy by which the p53-derived peptides are processed into Class I MHC, not by the steady-state level of p53. Tumour-associated antigen-specific CTL can mediate immune response of host against cancer in vivo. Targeting the missense mutation form can be a candidate of tumour antigen since cancer patients have antibodies against p53; although the frequency and clinical significance are matters of debate. Speetjens et al. reported clinical trials of a p53-specific synthetic long peptide (p53-SLP) vaccine for metastatic colorectal carcinoma. Ten patients were vaccinated with p53-SLP in phase I and II clinical trials. P53-specific T cells were isolated from the vaccination site, which were characterised as Th cells that displayed mixed T-helper 1 and 2 cytokine profile with different percentages of IFN- and IL-2 producing p53-specific T cells. Six of the patients showed strong proliferative p53-specific T cell response six months after the last vaccination.

This vaccine is viral vector-based. Menon et al., utilising the same concept of modality, performed a phase I/II clinical study in which end-stage colorectal cancer patients were vaccinated with a recombinant canarypox virus (ALVAC) encoding wild-type p53. Patients were immunised intravenously with an increasing dose of ALVAC-p53. The study reported that the vaccine was safe and capable of stimulating p53-specific Th1 (IFN-y) response in most patients. Fever was the only vaccine-associated adverse event reported. The conclusion of the authors was repeated immunisation would probably be needed for good clinical response. Antivector responses were also observed in all the patients.

Dendritic cell-based vaccines have also been tried, although it was reported that this model of vaccines was laborious to produce and restricted to individual patients. But it has an advantage of being highly efficient as APCs. Svane et al. tried this model in a phase II study in direct continuation of their phase I study. Only five out of the twenty-six patients completed all the planned immunisation scheduled as a result of rapid progression of disease or death. In most of the cases, an increase in the number of p53-specific CTLs was measured, but a decline at a late point after vaccination was observed.

Another concept is a peptide-based vaccine, in which patients are immunised with a single peptide epitope. However, the relatively poor immunogenicity of peptide epitopes means they need to be injected together with adjuvant. Rahma et al. utilised this approach when they compared subcutaneous wild-type p53 epitope vaccination with IV peptide-pulsed DC administered in twenty-one ovarian cancer patients, combined with IL-2 adjuvant, in a randomised phase II study. IL-2 administered resulted in direct induction of expanded Treg and in grade II/IV adverse event in both arms of the study, which were subsequently removed from the regimen of these patients. P53-specific T-cell was observed in about 70 per cent of the patients, irrespective of whether they received short peptide or peptide-pulsed DC.

Although limited success has been reported in these models, there is the need for further exploration. Also new vaccine strategies need to be tried; for example, Vermeij et al suggested that p53 vaccine can easily be combine with low-dose cyclophosphamide, anti-CTLA-4, chemotherapeutic regimens, or other tumour antigens as immunopotentiation treatment modalities. Future research should analyse whether addition of multiple antigen to p53 vaccine will elicit the required immune response. The best combination of therapy should be developed and identify those most likely to respond to combined anti-p53 therapy need to be established. Only then can we say we are nearer to déjà vu.

6.2.5   P53 Activation Pathway

The p53 protein family comprises p53, p63, and p73. They share the same protein structure and biological function. These are also isoforms of each gene and can heterodimerise each other. Since p53 is the most mutated in human cancer but not p63 or p73, the latter two could be used to step up their suppressive function in p53-defective cancers. Therefore, treatment that can increase p63 or p73 protein levels will deliver a p53-like function. One such compound that could activate the p53 pathway is NSC17627, which is an ellipticine derivative. Treatment of NSC176327 increased p53 target genes DR5 and DR1 expression. However, NSC176327 is less effective in inhibiting cell growth when p73 was knocked down in HCT16 p53 $^{-/-}$ cells, meaning p73 plays an important role in NSC176327-induced-p53-like activity.

Another compound with similar activity is reactivation of transcriptional reporter activity (RETRA), which was found to enhance p53 reporter activity in a mutated p53-dependent manner. Treatment with RETRA increased p73 expression and prevented the inhibition of p53 and p73, which produced a p53-like tumour suppression effect. However, the exact mechanism of how RETRA interferes with mutated p53 and p73 interaction is yet to be elucidated.

6.2.6 Replacement Gene Therapy

As outlined earlier, the function of p53 is lost in most cancers through mutation or loss of alleles. Therefore, restoring the function of p53 is a novel therapeutic strategy that can be achieved by replacing mutant gene with a functional wild-type version. A requirement for treatment of cancers with replacement gene therapies is the need for highly efficient delivery of the wild-type p53 into the tumour cells in vivo. Also, the function of p53 protein should be sufficiently expressed to mediate tumour suppression through either cell death or growth arrest, or by increasing the sensitivity to conventional antitumour agents. Low-level toxicity towards the normal cells should be considered in order not to develop certain pathologies. Gene delivery system is divided into two categories: viral and non-viral.

6.2.7 Other Molecules

Other compounds with various mechanism actions have been described. Benzodiapinedione (BDA) interacts with the p53-binding pocket of MDM2. This compound increases the p53 transcriptional activity, inhibits the proliferation of cancer cells in the wild-type p53-dependent manner, and synergises with doxorubicin to inhibit tumour cell growth in vivo and in vitro. Others include CP-31398 and CDB3. With the advent of improved screening and molecular technology, more compounds will be identified on a continuous basis due to the occurrence of new mutations.

6.3 Challenges and Future Perspective

Reactivating the function of p53 is an ideal strategy to control or conquer cancer. But because p53 malfunctioning varies among patients, novel and effective therapeutic interventions are still out of reach. A number of challenges would be encountered. Although it looks promising to target p53, it's not an ideal drug target because it is not a receptor or enzyme. Also, it is a heterotetrameric nuclear transcription factor, which is important to keep the genome stable and guard normal cells growth and physiological function.

However, advances accrued over the years are encouraging us to research further on developing tumour-specific p53 restoration therapies. We need to discover other roles than its function as a nuclear factor. The data so far show that cytoplasmic p53 can activate a transcriptional-independent apoptotic program. The next generation of p53-based therapies should therefore target this cytosolic function. Mammary stem cells with targeted p53 mutation have been reported. It showed the same properties as cancer stem cells. The reactivation of p53 restored the asymmetric cell division of cancer stem cells and induced tumour growth inhibition. Further studies are needed to elucidate the links between p53 function and cancer stem cells.

REFERENCES

Agarwal ML, et al (1998), The p53 network. J Biol Chem 273:1–4.

Aguiler F, et al (1993), Aflatoxin B1 induces the transversion of G to T in codon 249 of the p53 tumor suppressor gene in human hepatocytes. PNAS USA 90:8586–8590.

Aprea.com (2012), Aprea announces positive data from a clinical Phase I/II study with APR-246 | Aprea. [Online], accessed at: http://aprea.com/2012/aprea-announces-positive-data-from-a-clinical-phase-iii-study-with-apr-246/ (Accessed on 9th December 2015).

Antonia SJ, Mirza N, et al (2006), Combination of p53 cancer vaccine with chemotherapy in patients with extensive stage small cell lung cancer. Clinical Cancer Research 12:878–887.]

Bass IO, et al (1999), Clinical applications of detecting dysfunctional p53 tumor suppressor protein. Histol Histopathol 14: 279–284.

Benchimol S, Lamb P, et al (1985), Transformation associated p53 protein is encoded by a gene on human chromosome 17. Somat Cell Mol Genet 11: 505–510.

Boeckler FM, et al (2008), Targeted rescue of a destabilized mutant of p53 by an in silico screened drug. PNAS of USA 105: 0360–10365.

Bond GL, Hu W, et al (2004), A single nucleotide polymorphism in the MDM2 promoter attenuates the p53 tumor suppressor pathway and accelerates tumor formation in humans. Cell 119:591–602.

Bonnet MC, et al (2000), Recombinant viruses as a tool for therapeutic vaccination against human cancers. Immunology Letters 74:11–25.

Borrsesen AL, et al (1992), Papillomavirus, p53, and cervical cancer. Lancet 339: 1350-1351.

Bourdon JC, Fernandes K, et al (2005), p53 isoforms can regulate p53 transcriptional activity. Genes Dev 19: 2122–37.

Boyer SN, et al (1996), E7 protein of human papilloma virus-16 induces degradation of retinoblastoma protein through the ubiquitin-proteasome pathway. Cancer Res 56: 4620–4624.

Brehm A, et al (1998), Retinoblastoma protein recruits histone deacetylase to repress transcription. Nature 391: 597–601.

Brehm A, et al (1999), The E7 oncoprotein associates with Mi2 and histone deacetylase activity to promote cell growth. EMBO J 18: 2449–2458.

Bressac B, et al (1991), Selective G to T mutations of p53 gene in hepatocellular carcinoma from Southern Africa. Nature 350:429–431.

Brown CJ, et al (2009), Awakening guardian angels: Drugging the P53 pathway. Nature Reviews Cancer vol. 9: 862–873, 2009.

Buchman VL, et al (1988), A variation in the structure of the protein-coding region of the human p53 gene. Gene 70: 245–252.

Buetow K, et al (1992), Low frequency of p53 mutation observed in a diverse collection of primary hepatocellular carcinoma. PNAS USA 89:9622–9626.

Bui L, et al (2000), Transcription factor ZBP-89 cooperate with histone acetyltransferase p300 during butyrate activation of p21 waf1 transcription in human cells. J Biol Chem 275:30725–30733.

Bui L. Merchant JL (2002), ZBP-89 promotes growth arrest through stabilization of p5 molecule. Cell Biol 21: 4620–4683.

Busby-Eaile RM, et al (1992), Papillomavirus, p53, and cervical cancer. Lancet 339:1350.

Bykov VJ, et al (2002), Restoration of the tumor suppressor function to mutant p53 by a low-molecularweight compound. Nature Medicine 8, 282–288.

Caelles C, Helmberg A, Karin M (1994), P53-dependent apoptosis in the absence of transcriptional activation of p53-target genes. Nature 370: 220–223.

Cai DW, et al (1992), Stable expression of the wild-type p53 gene in human lung cancer cells after retrovirus-mediated gene transfer. Hum Gene Ther 4:617–624.

Challen C, et al (1992), Analysis of the p53 tumor-suppressor gene in hepatocellular carcinoma from Britain. Hepatology 16: 1362–1366.

Chen G, et al (2003), Mutation of p53 in recurrent hepatocellular carcinoma and its association with the expression of ZBP-89. Am J Pathol 162: 1823–1829.

Chen TM, Chen CA, et al (1992), The state of p53 in primary human cervical carcinoma and its effects in human papillomavirus immortalized human cervical cells. Oncogenes 8:1511–1518.

Cheok CF, et al (2011), Translating p53 into the clinic. Nature Reviews Clinical Oncology 8: 25–27.

Chiappori AA, et al (2010), INGN-225: a dendritic cell based p53 vaccine (Ad.p53-DC) in small cell lung cancer: observed association between immune response and enhanced chemotherapy effect. Expert Opinion on Biological Therapy 10: 983–991.

Chipuk JE, et al (2003), Pharmacologic activation of p53 elicits bax-dependent apoptosis in the absence of transcription. Cancer Cell 4:371–381.

Cho Y, et al (1994), Crystal structure of a p53 tumor suppressor –DNA complex: understanding tumorigenic mutations. Science 265:346–355.

Clayman GC, et al (1998), Adenovirus-mediated p53 gene transfer in patients with advanced recurrent head and neck squamous cell carcinoma. J Clin Oncol 16:2221–2232.

Crooks T, et al (1991), p53 mutation in HPV negative human cervical carcinoma cell lines. Oncogene 6:873–875.

Crooks T, et al (1992), Clonal p53 mutation in primary cervical cancer: association with human papillomavirus-negative tumors. Lancet 339:1070–1073.

Debuire B, et al (1995), Analysis of the p53 gene in European hepatocellular carcinoma and hepatoblastoma. Oncogene 8:2303–2306.

De Leo AB (1998), P53-based immunotherapy of cancer. Critical Reviews in Immunology 18: 29–35.

De Leo AB (2005), p53-based immunotherapy of cancer. Adv Otorhinolaryngol 62: 134–150.

de Mare A, et al (2008), Viral vector-based prime-boost immunization regimens: a possible involvement of T-cell competition. Gene Therapy 15:393–403.

Ding K, Lu Y, et al (2006), Structure-based design of spiro-oxindoles as potent, specific small-molecule inhibitors of the MDM2-p53 interaction. J Med Chem 49: 3432–3435.

Donehower LA (2002), Does p53 affect organism aging? J Cell Physiol 192: 23–33.

El-Deiry WS, et al (1993), WAF1, a potential mediator of p53 tumor suppression. Cell 75: 817–825.

Eastham JA, et al (2000), Suppression of primary tumor growth and the progression to metastasis with p53 adenovirus in human prostate cancer. J Urol 164:814–819.

Eliyahu D, Goldfinger N, et al (1998), Meth A fibrosarcoma cells express two transforming mutant p53 species. Oncogene 3: 313–321.

Enge M, et al (2009), MDM2-dependent downregulation of p21 and hnRNP K provides a switch between apoptosis and growth arrest induced by pharmacologically activated p53. Cancer Cell vol. 15: 171–183.

Fang WWX, Mazur W, et al (1994), High-efficiency gene transfer and high level expression of wild-type p53 in human lung cancer cells mediated by recombinant adenovirus. Cancer Gene Therapy 1: 5–13.

Figdor CG, et al (2004), Dendritic cell immunotherapy: mapping the way. Nature Medicine 10:475–480.

Finlay CA, et al (1988), Activating mutations for transformation by p53 produce a gene product that forms an hsc70-p53 complex with an altered half-life. Mol Cell Biol 8: 531–39.

Fridman JS, Lowe SW (2003), Control of apoptosis by p53. Oncogene 22: 9030–40.

Fujita M, et al (1992), Alteration of the p53 gene in human primary cervical carcinoma with and without human papillomavirus infection. Cancer Res 52: 5323–5328.

Fujiwara T, et al (1993), A retroviral wild-type p53 expression vector penetrates human lung cancer spheroids and inhibits growth by inducing apoptosis. Cancer Res 53:4129–4133.

Gallagher WM, Brown R (1999), p53-oriented cancer therapies. Ann Oncol 10: 139–150.

Gilboa E (2007), DC-based cancer vaccines. Journal of Clinical Investigation 117:1195–1203.

Grasberger BL, et al (2005), Discovery and cocrystal structure of benzodiazepinedione HDM2 antagonists that activate p53 in cells. J Med Chem 48: 909–912.

Greenblatt MS, et al (1994), Mutation in the p53 tumor suppressor gene: clues to cancer etiology and molecular pathogenesis. Cancer Res 54: 4855–4878.

Green DR and Kroemer G (2009), Cytoplasmic functions of the tumour suppressor p53. Nature 458: 1127–1130.

Guo J and Xin H (2006), Chinese gene therapy. Splicing out the West?" Science, v314; 232–1235.

Gu L, Zhu N, et al (2008), MDM2 antagonist nutlin-3 is a potent inducer of apoptosis in pediatric acute lymphoblastic leukemia cells with wild-type p53 and overexpression of MDM2, Leukemia; 22:730–739

Hainaut P, Hollstein M (2000), p53 and human cancer: the first thousand mutations. Adv Cancer Res 77:81–137.

Hainaut P, et al (1995), Temperature sensitivity for conformation is an intrinsic property of wild type p53. Br J Cancer 71:227–231.

Hall PA, Lane DP (1997), Tumor suppressors: a developing role for p53? Curr Biol 7: 144–47.

Hernandez-Boussard T, et al (1999), IARC p53 mutation database: a rational database to compile and analyze p53 mutations in human tumors and cell line. International Agency for Research in Cancers, Hum Mutation 14:1–8.

Herrin V, Behrens RJ, et al (2003), Wild-type p53 peptide vaccine can generate a specific immune response in low burden ovarian adenocarcinoma. In: Proceedings of the American Society of Clinical Oncology Annual Meeting.

Herrin V, Achtar M, et al (2007), A randomized phase II p53 vaccine trial comparing subcutaneous direct administration with intravenous peptide-pulsed dendritic cells in high risk ovarian cancer patients. In: Proceedings of the American Society of Clinical Oncology Annual Meeting.

Hollstein MC, Sidransky D, et al (1991), p53 mutations in human cancers. Science 253: 49–53.

Hollstein MC, et al (1993), p53 mutations and aflatoxin B1 exposure in hepatocellular carcinoma in patients from Thailand. Int J Cancer 53:51–53.

Hung CF, et al (2008), Antigen-specific immunotherapy of cervical and ovarian cancer. Immunological Reviews 222:43–69.

Inoue A, et al (2000), Administration of wild-type p53 adenoviral vector synergistically enhances the cytotoxicity of anti-cancer drugs in human lung cancer cells irrespective of the status of p53 gene. Cancer Lett 157:105–112.

Issaeva N, et al (2004), Small molecule RITA binds to p53, blocks p53-HDM-2 interaction and activates p53 function in tumors. Nature Medicine 2: 1321–1328.

Itahana K, Dimri G, Campisi J (2001), Regulation of cellular senescence by p5. Eur J Biochem 268: 2784–91.

Jin S, Levine AJ (2001), The p53 functional circuit. J Cell Sci 114: 4139–4140.

Kanodia S,et al (2008), Recent advances in strategies for immunotherapy of human papillomavirus-induced lesions. International Journal of Cancer 122:247–259.

Kirk GD, et al (2000), Ser-249 p53 mutations in plasma DNA of patients with hepatocellular carcinoma from the Gambia. J Natl Cancer Inst 92:148–153.

Koblish HK, Zhao S,et al (2006), Benzodiazepinedione inhibitors of the Hdm2:p53 complex suppress human tumor cell proliferation *in vitro* and sensitize tumors to doxorubicin *in vivo*. Mol Cancer Ther 5:160–169.

Komarov PG, et al (1999), A chemical inhibitor of p53 that protects mice from the side effects of cancer therapy. Science 285:1733– 1737.

Kramer DL, et al (1999), Polyamine analogue induction of p53-p21 WAF1/ CIP!-Rb pathway and G1 arrest in human melanoma cells. Cancer Res 59: 1278–1286.

Kravchenko JE, et al (2008), Small-molecule RETRA suppresses mutant p53-bearing cancer cells through a p73-dependent salvage pathway. PNAS USA 105:6302–6307.

Lain S, et al (2008), Discovery, in vivo activity, and mechanism of action of a small-molecule p53 activator. Cancer Cell 13: 454–463

Lamb P, Crawford L (1986), Characterization of the human p53 gene. Mol Cell Biol 6: 1379–1385.

Lambert J, et al (2009), PRIMA-1 reactivates mutant p53 by covalent binding to the core domain. Cancer Cell 15: 376–388.

Lane DP (1992), Cancer p53 guardian of the genome. Nature 358:15–16.

Lane DP, Hall PA (1997), MDM2-arbiter of p53's destruction. Trends Biochem Sci 22: 372–374.

Laurie NA, Donovan SL, et al (2006), Inactivation of the p53 pathway in retinoblastoma. Nature 444:61–66.

Leffers N, Lambeck AJA, et al (2009), Immunization with a P53 synthetic long peptide vaccine induces P53-specific immune responses in ovarian cancer patients, a phase II trial. International Journal of Cancer 125:2104–2113.

Levine AJ (1997), p53, the cellular gate keeper for growth and division. Cell 88:323–331

Lewis EJ (2015), PRIMA-1 as a cancer therapy restoring mutant p53: A review. Bioscience Horizons 82015:1–5.

Levine AJ (1997), p53, the cellular gatekeeper for growth and division. Cell 88: 323–31.

Levine AJ, Momand J, Finlay CA (1991), The p53 tumour suppressor gene. Nature 351: 453–456.

Liang XH, et al (1993), Co-localization of the tumor suppressor protein p53 and human papillomavirus E6 protein in cervical carcinoma cell line. Oncogene 8:2645–2652.

Li D, et al (1992), p53 gene mutation spectrum in hepatocellular carcinoma. Carcinogenesis 14: 169–173.

Linares LK, Hengstermann A, et al (2003), HdmX stimulates Hdm2-mediated ubiquitination and degradation of p53, PNAS USA; 100:12009–12014

Liu Y, Bodmer WF (2005), Analysis of p53 mutations and their expression in 56 colorectal cancer cell lines, PNAS USA; 103:976–981.

Liu Y, et al (2009), P53 regulates hematopoietic stem cell quiescence. Cell Stem Cell 4: 37–48.

Lowe SW (1995), Cancer therapy and p53. Curr Opin Oncol 7: 347–355.

Lu C, Wang W, El-Deiry WS (2008), Non-genotoxic anti-neoplastic effects of ellipticine derivative NSC176327 in p53-deficient human colon carcinoma cells involve stimulation of p73. Cancer Biol Ther 7:2039–2046.

Martins CP, et al (2006), Modeling the therapeutic efficacy of p53 restoration in tumors. Cell 127: 323–1334.

Matlashewski G, et al (1984), Isolation and characterization of a human p53 cDNA clone: Expression of the human p53 gene. EMBO J; 3: 3257–62.

Medrano S, Scrable H (2005), Maintaining appearances--the role of p53 in adult neurogenesis. Biochem Biophys Res Commun 331: 828–33.

Meletis K, et al (2006), P53 suppresses the self-renewal of adult neural stem cells, Development 133: 63–369.

Menon AG, et al (2003), Safety of intravenous administration of a canarypox virus encoding the human wild-type p53 gene in colorectal cancer patients. Cancer Gene Therapy 10:509–517.

McBride OW, Merry D, Givol D (1986), The gene for human p53 cellular tumor antigen is located on chromosome 17 short arm (17pl3). PNAS USA 83: 130–34.

Michalovitz D, Halevy O, Oren M (1990), Conditional inhibition of transformation and of cell proliferation by a temperature-sensitive mutant of p53. Cell 62: 671–680.

Momand J, Zambetti GP, et al (1992), The mdm-2 oncogene product forms a complex with the p53 protein and inhibits p53-mediated transactivation. Cell 69:1237–45.

Mowat M, et al (1985), Rearrangements of the cellular p53 gene in erythroleukaemic cells transformed by Friend virus. Nature 314: 633–36.

Mutka SC, Yang WQ, et al (2009), Identification of nuclear export inhibitors with potent anticancer activity in vivo. Cancer Res 69:510–7.

Naito M, et al (1992), Detection of p53 gene mutation in human ovarian and endometrial cancers by polymerase chain reaction- single strand confirmation polymorphism analysis. Jpn J Cancer Res 83: 1030–1036.

Oder T, et al (1992), p53 gene mutation spectrum in hepatocellular carcinoma. Cancer Res 52: 6358–6364.

Offringe R, et al (2000), p53: a potential target antigen for immunotherapy of cancer. Ann NY Acad Scin 910:223–233.

Oliner JD, et al (1992), Amplification of a gene encoding a p53-associated protein in human sarcomas. Nature 358: 80–83.

Olivier M, Hussain SP, et al (2004), TP53 mutation spectra and load: a tool for generating hypotheses on the etiology of cancer. IARC Sci Publ 157: 247–70.

Oren M (2003), Decision making by p53: life, death and cancer. Cell Death Differ 10: 431–42.

Rahma O, Achtar M, et al (2010), Comparable effect of p53 peptide vaccine in adjuvant or pulsed on denritic cells in ovarian cancer patients: a gynecologic oncology group study, In: Proceedings of the American Society of Clinical Oncology Annual Meeting.

Ramakrishnan R, Antonia S, Gabrilovich DI (2008), Combined modality immunotherapy and chemotherapy: A new perspective. Cancer Immunology, Immunotherapy 57:1523–1529.

Ramakrishnan R, et al (2010), Chemotherapy enhances tumor cell susceptibility to CTL-mediated killing during cancer immunotherapy in mice. Journal of Clinical Investigation 120:1111–1124.

Reifenberger G, et al (1993), Amplification and overexpression of the MDM2 gene in a subset of human malignant gliomas without p53 mutations. Cancer Research 53: 2736–2739.

Ringshausen I, et al (2006), Mdm2 is critically and continuously required to suppress lethal p53 activity in vivo. Cancer Cell 10: 501–514.

Rippin TM, et al (2002), Characterization of the p53-rescue drug CP-31398 in vitro and in living cells. Oncogene 21: 2119–2129.

Rogel A, Popliker M, et al (1985), p53 cellular tumor antigen: analysis of mRNA levels in normal adult tissues, embryos, and tumors. Mol Cell Biol 5: 2851–55.

Rosenberg SA (1999), A new era for cancer immunotherapy based on the genes that encodes cancer antigens. Immunity 10: 281–287.

Rosenberg SA, Dudley ME (2004), Cancer regression in patients with metastatic melanoma after the transfer of autologous antitumor lymphocytes. PNAS USA 101: 14639–14645.

Roth JA, et al (1996), Retrovirus mediated wild-type p53 gene transfer to tumors of patients with lung cancer. Nature Medicine 2: 985–991.

Sangrajrang S, et al (2003), Serum p53 antibodies in patients with lung cancer: correlation with clinicopathologic features and smoking. Lung Cancer 39: 297–30.

Scheffner M, et al (1990), The E6 oncoprotein encoded by human papillomavirus type 16 and 18 promotes the degradation of p53, Cell; 63:1129–1136.

Schlom J, Arlen PM, Gulley JL (2007), Cancer vaccines: moving beyond current paradigms, Clinical Cancer Research; 1:3776–3782.

Shangary S, et al (2008), Temporal activation of p53 by a specific MDM2 inhibitor is selectively toxic to tumors and leads to complete tumor growth inhibition. PNAS of USA 105: 3933–3938.

Shimada H, Matsubara H, et al (2006), Phase I/II adenoviral p53 gene therapy for chemoradiation resistant advanced esophageal squamous cell carcinoma. Cancer Science 97:554–561.

Soussi T (2000), The p53 tumor suppressor gene: from molecular biology to clinical investigation. Ann NY Acad Sci 910:121–137.

Soussi T (2000), p53 antibodies in the sera of patients with various types of cancer: a review. Cancer Res 60: 1777–1788.

Speetjens FM, et al (2009), Induction of p53-specific immunity by a p53 synthetic long peptide vaccine in patients treated for metastatic colorectal cancer. Clinical Cancer Research 15: 1086–1095.

Srivastava S, et al (1992), The status of the p53 gene in human papillomavirus positive or negative cervical cancer cell lines. Carcinogenesis 13:1273–1275.

Stewart N, Hicks GG, et al. (1995), Evidence for a second cell cycle block at G2/M by p53. Oncogene 10: 109–15.

Sun SH, Zheng M, et al (2008), A small molecule that disruptsMdm2-p53 binding activates p53, induces apoptosis, and sensitizes lung cancer cells to chemotherapy. Cancer Biol Ther 7:845–852.

Suzuki K, Matsubara H (2011), Recent advances in p53 research and cancer treatment. J of Biomed and Biotech 2011, Article ID 978312; doi: 10.1155/2011/978312.

Svane IM, et al (2004), Vaccination with p53-peptide-pulsed dendritic cells, of patients with advanced breast cancer: report from a phase I study. Cancer Immunology, Immunotherapy 53:633–641.

Svane IM, Pedersen AE, et al (2007), Vaccination with p53 peptide-pulsed dendritic cells is associated with disease stabilization in patients with p53 expressing advanced breast cancer; monitoring of serum YKL-40 and IL-6 as response biomarkers. Cancer Immunology, Immunotherapy 56:1485–1499.

Tacken PJ, et al (2007), Dendritic-cell immunotherapy: from ex vivo loading to in vivo targeting. Nature Reviews Immunology 7:790–802.

Teramoto T, et al (1994), p53 gene abnormalities are closely related to hepatoviral infections and occur at a late stage of hepatocarcinogesis. Cancer Res 54:231–235.

Tokino T, Nakamura Y (2000), The role of p53 target-genes in human cancer. Crit Rev Oncol Hematol 33:1–6.

Tsuda H, Hiroheshi S (1992), Frequent occurrence of p53 gene mutation in uterine cancers at advances clinical stage and with aggressive histological phenotype. Jpn J Cancer Res 83: 1184–1191.

Van der Burg SH, et al (2002), Induction of p53-specific immune responses in colorectal cancer patients receiving a recombinant ALVAC-p53 candidate vaccine. Clinical Cancer Research 8:1019–1027.

Vassilev LT, et al (2004), In vivo activation of the p53 pathway by small-molecule antagonists of MDM2. Science. 303: 844–848.

Ventura A, et al (2007), Restoration of p53 function leads to tumour regression in vivo. Nature 445: 661–665.

Vermeij R, Leffer N, et al (2011), Immunological and clinical effects of vaccines p53 overexpressing malignancies. J Biomed Biotech 2011: 7021146.

Vogelstein B, Kinzler KW (1992), p53 function and dysfunction. Cell 70: 523–26.

Vogelstein B, Lane D, Levine AJ (2000), Surfing the p53 network. Nature 408: 307–10.

Vojtesek B, et al (1995), Absence of p53 autoantibodies in a significant proportion of breast cancer patients. British Journal of Cancer 71: 1253–1256.

Wang W, El-Deiry WS (2008), Restoration of p53 to limit tumor growth. Curr Opin Oncol 20:90–96.

Zeimet AC, et al (2000), New insight into p53 regulation and gene therapy for cancer. Biochem Pharmacol 60:1153–1163.

Zhang L, et al (2008), Differential impairment of regulatory T cells rather than effector T cells by paclitaxel-based chemotherapy. Clinical Immunology 129:219–229.

Zhang WW, et al (1994), High-efficiency gene transfer and high level expression of wild-type p53 in human lung cancer cells mediated by recombinant adenovirus. Cancer Gene Therapy 1: 5–13.

Zou W (2005), Immunosuppressive networks in the tumour environment and their therapeutic relevance. Nature Reviews Cancer 5:263–274.

CHAPTER 7

HPV ONCOPROTEIN AND CARCINOMA

7.0   Introduction

Human papillomavirus (HPV) are small epitheliotropic double-stranded DNA viruses that are involved in the aetiology of several human cancers. The HPV family includes about two hundred categories, of which forty have been isolated from the genital tract. HPVs are divided into two types based on their clinical association: high- and low-risk. The high-risk HPVs are associated with the development of cervical carcinoma in persisted infected females. It is estimated that about 90 per cent of all cervical carcinomas are associated to high-risk HPVs. Of these, HPV-16,-18, and -31 are the most prevalent and commonly linked to lesions that may progress to high-grade intraepithelial neoplasia and ultimately carcinoma. The others, which include HPV-6 and -11, are low-risk and associated with benign lesions that rarely progress to cancer. HPV-16 and -18 account for approximately 50 per cent of all cases of cervical cancer, which is responsible for one-fifth of all cancer-associated deaths among women diagnosed each year, making it the second most common cancer among women worldwide. Some of the low risk may be transmitted via sexual contact to cause genital condyloma. All the types of the virus share a common genomic structure (figure 1), which encodes eight proteins. These include six early E1, E2, E4, E5, E6, and E7, and two late proteins. E5, E6, and E7 oncoproteins of the high-risk strains are considered anti-apoptotic oncoproteins, which plays a significant role in malignant transformation. In addition, E2 and E7 are considered proapoptotic proteins.

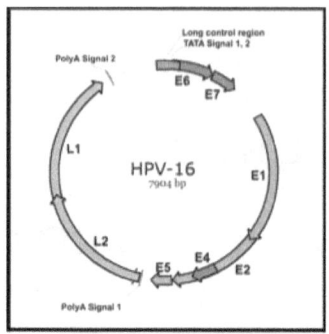

Figure 7. 1: Genomic structure of HPV (Source: National Institute of Cancer, 2013)

The importance placed on our understanding of HPV as a contributor to malignant progression resulted in the 2008 Nobel Prize going to Harald zur Hausen for his discovery that high-risk HPV types are associated with the development of cervical cancer. Not everyone who is infected with high-risk HPVs develops cancer. Additional genetic alteration is required for the progression to malignancy. The HPV oncoproteins E5, E6, and E7 are the primary viral factors associated with the initiation and progression of cervical cancer. They act by overcoming the negative growth instability which is the hallmark of HPV-associated cancers. This chapter will review some of the latest works on the pathogenesis of HPV infection and the role played by these oncoproteins. The life cycle of the virus will be looked at, and other important themes such as therapeutic and vaccine intervention will be analysed.

7.1   HPV Life Cycle

HPV has an unusual life cycle (figure 7.2). In most viruses, the target cell is infected, which results in the production of progeny virus from the same infected cells. In HPV infection, the production of new virions occurs only after cell differentiation. HPV infects cells found in the basal layer of the stratified squamous epithelial. After infection, the viral genomes are established as an extrachromosomal entity or episome. The HPV genomes are small and do not encode polymerase or other enzymes required for viral replication. The virus therefore depends on the host cell replication proteins to mediate viral DNA synthesis. In HPV infection, the suprabasal cells remain active in the cell cycle as they undergo the process of differentiation with a subset of cells entering the S phase to replicate the HPV genome. This process is referred to as amplification. It is followed by capsid protein synthesis, virus assembly, and release.

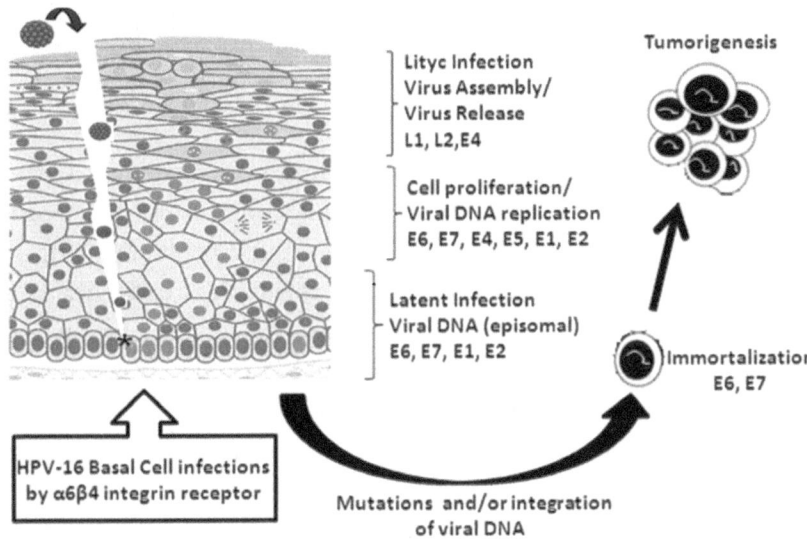

Figure 7.2: A typical life cycle of HPV

The proliferative ability of HPVs is controlled by a number of cellular factors, the most important being the members of retinoblastoma (Rb) family. HPV E7 proteins bind to the Rb family and target them for degradation. This leads to the release and activation of E7F transcription factors that promote the expression of S phase genes. It has been shown that E7 proteins from all HPV types bind Rb family members, but the high-risk E7 protein binds with much higher affinity. Efficient binding of Rb by E7 can result in inhibited cell growth and apoptosis through a p53-dependent pathway. By their combined action, high-risk E6 and E7 proteins target these cell cycle regulators to maintain S phases, which are competent in differentiation cells and also leads to abrogation of many cell cycle checkpoints. In cells with persistently HPV infection, there is an accumulation of cellular mutations over time, with a consequent progression to cancer. The high-risk E5 protein combines rather fatally with E6 and E7 proteins to promote hyperproliferative of infected cells, which might facilitate the progression to cancer. From infection to the initiation of cancer, a series of molecular events do take place. The failure of the immune system to clear persistent HPV infection can result in the development of cervical cancer years after harbouring the virus. In precancerous lesions, most of the viral genome persists in episomal form, while in the high-grade lesions, the genomes are integrated into the host chromosome, with the fragile sites the most often area of integration.

It has been suggested that integration might contribute to cancer progression. Viral E2 protein in the lesions containing HPV reduces early gene expression. Integration of viral DNA in most cases disrupts E2 expression, which leads to the deregulated expression of early viral genes, including E6 and E7. There is also increased capacity for proliferation, which is a crucial step in cancer progression. E6 and E7 are essential for maintaining the transformed phenotype. Kadaja et al. described an important step in HPV carcinogenesis: coexistence of HPV episomes and integrated copies. Replication of integrated sites of origin leads to activation of DNA repair and recombination system. This increases the chance of acquiring cellular mutation and increases genomic instability, thereby increasing the opportunity for cancer progression. The various mechanisms associated with cancer development in HPV infection will be dealt with later.

7.2   E6, E5, and E7 Oncoproteins

7.2.1   E6 Oncoprotein

The HPV E6 oncoprotein is a relatively small protein which was long recognised as a potent oncogene and ultimately associated with events that result in the malignant conversion of virally infected cells. Analysis of HPV-16 E6 oncoprotein showed that it's made of 150 amino acids, containing two CX2C-X29-CX2C zinc-like fingers, which are joined by interdomain linkers of 36 amino acids. In the high-risk E6 genes, shortened E6 can inactivate the function of full-length E6 by binding to the interface of the C- and N- terminal halves of E6 proteins. E6 oncoprotein does not have any enzymatic activities, but most of its major activities are triggered by protein–protein interactions. The first protein that interacts with E6 is E6-associated protein (E6AP), which is an ubiquitin ligase. E6, E6AP, and target protein form a complex that results in the ubiquitination of the target protein, with a subsequent proteasome-mediated degradation. One of the main targets of E6 is p53, a key signaling coordinator in the cell, after genotoxic or cytotoxic stress. E6 protein binds to p53 with the assistance of E6AP and stops p53 from inducing apoptosis by targeting it for degradation through ubiquitin-pathway. E6 oncoprotein is involved in two pathways that are associated with apoptosis: p53 inactivation and blocking apoptosis. In the first instance, p53 inactivation may trigger the E6-induced apoptosis inhibition. The most important mechanism for p53 activation by high-risk HPV is induction of p53 degradation via the ubiquitin-proteasome pathway. Additionally, E6 protein in high-risk HPV can inhibit p53 activation by blocking the alternative reading frame p14 pathway and by interacting with a histone acetyltransferase, hAD3. Secondly, inhibition of apoptosis maybe triggered by E6 oncoprotein. It has been found that p53-independent apoptosis is also capable of eliminating abnormal cells, while E6 is capable of blocking apoptosis in cells and mice lacking p53. There are two major apoptotic pathways: intrinsic and extrinsic pathways. E6 is able to disturb these pathways and prevent cell death under endogenous and exogenous stress.

7.2.2   E5 Proteins

E5 is the smallest among HPV oncoproteins, and its ORF has been classified into four different groups based on the different clinical manifestations, especially with the potential for oncogenesis: alpha, beta, gamma, and delta. In HPV-16, E5 is encoded by eighty-three amino acids. The E5 ORF is absent in the genome of many HPVs, which indicates that the protein is not essential for the life cycle of these viruses, but it gives an added value to favour infection and transformation. HPVs are thought to use leaky ribosome-scanning mechanisms to translate protein from polycistromic mRNA.

Little E5 protein is mostly likely to be synthesised from these transcripts. However, on epithelial-cell differentiation, E5 is expressed on the second ORFs of late transcript. The suggestion is E5 is likely synthesised mostly in differentiating suprabasal epithelial cells. Detection of E5 protein is difficult due to its extreme hydrophobicity, membrane localisation and low levels of expression. Chang et al. reported that since HR E5 is able to integrate into the human genome during malignant progression, the E5 gene is rarely found in cervical tumours. HPV E5 inhibits death receptor-mediated apoptosis in human keratinocytes and is capable of downregulating the total amount of Fas receptor and decreasing Fas location. It also alters the formation of DISC induced by TRAIL. This means E5 can impair Fas ligand (FasL) and TRAIL-mediated apoptosis.

HPV-16 E5 protein has been found to suppress three main proteins in the ER stress pathway. These include cyclooxygenase-2 (COX-2), X-box binding protein-1 (XBP-1), and inositol-requiring enzyme-1α (IRE1α). The downregulation of these is beneficial for viral replication and persistence. In cervical cancer cells, EP4 protein can be activated by HPV-16 E5, which activates protein-kinase A, responsible for the anti-apoptotic effect. Additionally, activating EP4 can enhance the expression of VEGF, with a resultant tumour immortalisation in cervical carcinoma. The transforming activity of E5 has been described in a number of cell types and assays.

7.2.3 E7 Proteins

Oncoprotein E7 is a small acidic polypeptide encoding for about 100 amino acids. It shares functional similarities with other viral oncoproteins, especially adenoviral E1A and SV40 large T antigen, both in primary sequence, transactivation, and transformation properties. Based on its amino acid sequence, E7 can be separated into three conserve regions, just like in adenovirus E1A as CR1, CR2, and CR3. The CR2 and CR3 regions of HPV E7 share the same sequence with the corresponding regions of adenovirus E1A and SV40 large T antigen, including a strictly conserved CXCEX motif, which mediates high affinity for pRb. The CR3 region of E7 is made of two CXXC motifs, which are separated by twenty-nine or thirty residues, forming a zinc binding domain. Studies have shown that HPV C3 mediates protein dimerisation and mediates direct interaction with several E7-interacting proteins. The E7 CR3 region can also mediate inactivation of the cyclin-dependent kinase inhibitors p27 and p21 as well as several transcription factors that are implicated in HPV-associated oncogenesis, including TATA-box-binding protein (TBP), the transcription factor E2F, and p300/CBP-associated factor (P/CAF).

The HPV-16 E7 oncoprotein induces p53-depedent and independent apoptosis. It mediates cell transformation in part by binding to the human pRb tumour suppressor protein and E2F transcription factor, which leads to dissociation of pRb from E2F transcription factor and the prevents cell progression to S phase of the cell cycle. This activity is mediated by Leu-X-Cys-X-Glu (LXCXE) motif and CR3 zinc binding domain of the E7 protein. E7 leads to the anti-apoptotic pRb degradation through a mechanism involving association with and reprogramming of cullin 2 ubiquitin ligase complex. It has been shown that the C-terminal of E7 oncoprotein is made of a low-affinity pRb binding site, which interacts with pRb.

Other pathways have been identified that trigger apoptosis with the involvement of E7 oncoprotein. For example, E7 protein activates apoptosis in NIH3T3 cell through a conserved LXCXE motif in the second chromodomain (CD2) of the E7 structure binding to pRb. Another pathway involves E7 and p21 forming a complex which activates Cathepsin B, a known apoptotic mediator. Another pathway involves E7 oncoprotein inducing TRAIL and TNF-α-induced apoptosis in primary human keratinocytes. In addition, it has been reported that E7-induced apoptosis is associated with interaction between E7 and E2F1. The complex is able to trigger E2F1-driven transcription, contributing to increased apoptosis. Other mechanisms of inhibiting apoptosis and cytokine-mediated cell death by HPV E7 oncoprotein have been described. It was reported that HPV-16 E7 interacts with and abrogates the growth-inhibitory activities of cyclin-dependent kinase inhibitor (CKI) sp21 and CKI sp27 as well as abrogating TGF-β-mediated growth inhibition.

Another study reported that HPV-16 E7 protein inhibits TNF-α-mediated apoptosis in normal human fibroblasts by upregulating the expression of inhibition of apoptosis (IAP) protein, c-IAP2. This is aided by suppression of caspase 8 activation. Siva-1 is a proapoptotic cellular factor capable of binding to the anti-apoptotic protein Bcl-X2. When HPV E7 interferes with the binding of Siva-1 and Bcl-XL, the released Bcl-XL is able to exert its anti-apoptotic effect. Finally, it has been reported that HR or LR HPV are able to interact with the 600kDa pRb-associated factor, p600. Therefore, the conjugation of E7 and p600 can protect the detached cells from apoptosis, thereby contributing to viral transformation.

7.3   Biomarkers of HPV-Associated Carcinomas

A biomarker is a substance or process that is indicative of the presence of cancer in the body. The US National Institute of Health's (NIH's) Working Group and the Biomarkers Consortium define biomarkers as "a characteristic that is objectively measured as an indicator of normal biologic process, pathogen process, or a pharmacological response to a therapeutic intervention." The NIH's National Cancer Institute (NCI) describes biomarkers as "biological molecules found in blood, other body fluids, or tissues that is a sign of a normal process, or a condition or disease." The World Health Organization defines biomarkers as any substance, structure, or process that can be measured in the body or its product and influences or predicts the incidence of outcome or disease. Biomarkers have also been defined as a measureable phenotypic parameter that characterises an organ, state of health or disease, or a response to a particular therapeutic intervention. Biomarkers can also be defined based on physical, chemical, or biological entities which are accessible in the body matrices and measureable in bodily fluids or cells. Biomarkers are sometimes referred to as molecular markers and signature molecules.

A cancer biomarker may be a molecule secreted by a tumour or a specific response of the body to the presence of cancer. Genetic, epigenetic, proteomic, glycomic, and imaging biomarkers can be used for cancer diagnosis, prognosis, and epidemiology. Ideally, such biomarkers should be assayed in non-invasively collected biofluids such as blood or serum. Several biomarkers have been identified in HPV infections and carcinomas which identify specific stages in the natural history of HPV infection and cancer progression. They include the presence of viral nucleic acids and viral proteins, and the alteration of cellular factors induced by viral oncoproteins.

In cervical cancer, numerous assays have been developed to detect nucleic acids of oncogenic and non-oncogenic HPVs in cervical cancer. The advantage of using HPV testing is that it is highly sensitive, with high predictive values, due to the fact that the absence of carcinogenic HPV indicates an extreme risk of CIN3 or cancer, and a long protection compared to cytology because the risk of CINs and cancer remains very low up to five years after a negative HPV test.

There is one concern: the low specificity of HPV assays because they cannot discriminate between transient and persistent HPV infections. Nevertheless, data from a number of studies show that high-risk HPV testing is highly sensitive and specific for detecting CIN2. A number of HPV assays are grouped into two categories based on their ability to identify a pool of HR HPVs types with or without genotyping of the most common high risks ones, or to detect a broad spectrum of oncogenic and nononcogenic HPVs along with individual genotyping. The assays available include HPV DNA assays, of which four have been approved by FDA: 1. Hybrid capture 2 (Qiagen), detecting thirteen HR HPVs; 2. Cervista HPV (Hologic), targeting 14 HR HPVs; 3. Cervista HPV 16/18 (Hologic), specifically designed for identifying HPV 16 and 18; and 4. Cobes 480 HPV (Roche Diagnostics), targeting 14 HR HPVs. HPV RNAs assays have been developed that detect viral mRNA encoded by E6 and E7 oncoproteins. Four assays are currently available for detecting HPV E6 mRNA in cervical samples: APTIMA (Gen Probe) and Onco Tech (Incell DX) assays are based on RT and PCR techniques to detect E6/E7 from 14 and 13 HR HPV genotypes, respectively. PreTech HPV-Proofer (Norchip) and NucliSENS EasQ (Biomerieux) both rely on nucleic acid sequence-based amplification (NASBA). They are able to detect E6/E7 transcripts from five of the most common HR HPV types in cervical carcinoma (HPV 16, 18, 31, 33, and 45).

Two HPV protein assays are also available for detecting protein levels in cervical cell exfoliates. These are OncoE6 (Arbor Vita Corporation) and Cyto Active (Cytoimmune Diagnostic). OncoE6 is used to identify the E6 protein encoded by HPV 16, 18, and 45. The Cyto Active assays are designed to measure the loss of L1 expression, which is a predictive marker of progressive lesions. In addition to these biomarkers, cellular biomarkers have been developed such as P161NK4a assay, which is an important tool for the improvement of the diagnostic accuracy, reliability, and quality in histopathology analysis of cervical lesion; BD Pro ExCTM marker assay has a higher sensitivity for detecting women with low SIL than women with high SIL.

The E6 protein promotes transcription of telomerase reverse transcriptase (TERT), which stabilises and repairs repeated DNA sequences at the telomere end of chromosomes. Gains of chromosome 3q, which contains the sequence for the telomerase RNA component (TERC), and gains of chromosome 5p containing TERT gene are associated with CIN2 or worse in cervical tissue biopsies. Therefore, evaluation of chromosome 3q and 5p with FISH and multiplex ligation-dependent probe amplification (MLPA) may be a useful marker for diagnosing progressing lesions. But the analysis of TERT and TERC copy number increase will be difficult in cytology samples as a result of the presence of predominantly normal cells in the samples.

One novel indicator of HPV carcinogenesis is miRNAs. Several miRNAs such as miR-9, miR-21, miR-27a, and miR-127 have been described in the dysregulation of cervical cancer and other HPV-associated carcinomas. Riberio et al. reported that in 114 women with different cervical lesions, miR34a expression seems to correlate with invasive cervical cancer, while miR-125b expression significantly changed with different cervical lesions. Li et al. also reported that among 171 women with CIN, miR-218 levels were lower in patients with HR HPV than those with LR or intermediate

HPVs. Wang et al. showed that the expression of miR-375 in 170 cervical cancer tissues significantly reduced compared to 68 normal tissues, suggesting that downregulation of miR-375 could be associated with cervical cancer. Malta et al. showed that Let-7c, a microRNAs precursor, significantly changed in exfoliated cervical cells from women with cervical intraepithelial lesions. Other molecules that have the potential to function as predictive markers of cervical cancer and other HPV-associated carcinomas include Yes-associated protein (YAP), Laminin receptor, and E-cadherin. As more alternative biomarkers result in better diagnosis and management of HPV-associated carcinomas, basic and clinical studies will identify and optimise other available biomarkers.

7.4 Anti-Oncoprotein Agents, Novel Approaches

Current treatment strategy for HPV-associated carcinomas involves use of radiotherapy with cisplatin-based chemotherapy and surgery, while a limited success has been achieved with immune modulators such as IFN. In addition, prevention of HPV infection by vaccination and challenge of established HPV infections by immune therapy is still being investigated. With the radiotherapy, it involves radiotherapy with cisplatin-based chemotherapy. Statistically, about half a million mortality associated with cervical cancer are reported yearly, and a much higher number of patients are exposed to noninvasive disease or genital warts. We can conclude from this data that current treatment of HPV-associated carcinomas is not adequate. Development of alternative treatment options is needed. With advances in molecular biology, it is now possible to target specific HPV proteins involved in virulence.

A number of novel ideas have been investigated. This can be done by optimisation some existing drugs or identifying new compounds that can inhibit proteins in the virus's life cycle. With the former, a number of novel optimisation studies have been undertaken. For example, Bortezomib is a proteasome inhibitor, while suberoylanilide hydroxamic acid (SAHA, also called Vorinostat) is a histone deacetylase inhibitor which is recognised as a potential chemotherapeutic drug. Both drugs have been approved by the FDA for the treatment of cutaneous T cell lymphoma and multiple myeloma/mantle cell lymphoma. In a study, Huang et al. showed that combination of bortezomib and SAHA elicited potent antitumour TC1-tumour-bearing mice, which led to tumour-specific immunity by antigen-specific CD8+ T cells than treatment with either drug alone. They concluded that their study could serve as the foundation for future application of both drugs for the treatment of cervical cancer.

Earlier, Lin et al. examined the effects of SAHA with bortezomib on tumour cell apoptosis. They found that the combination elicited synergistic killing of HPV-expressing cervical cancer cell lines, and the combination treatment diminished tumour growth of HeLa xenografts more effectively than either drug alone. In a systematic review study, this author concluded the IM IFN in combination with 5 per cent imiquimod has an effect on HPV-associated cervical cancer, but study is needed to evaluate this combination in clinical trial. Addition of platin-based chemotherapy to radiotherapy has increased five-year survival of advanced stage of cervical cancer patients. However, cases of treatment failure have been reported. One of the factors associated with treatment failure is the ability of the tumour cells to repair chemotherapy-induced DNA damage.

Therefore, a novel strategy has been suggested to improve the therapeutic effect: sensitisation of tumour cells for chemotherapy via inhibition of the DNA damage response (DDR). Cervical carcinogenesis involving HPV infection inactivates part of the DNA damage response. Therefore, HPV-mediated partial inactivation of DDR is a novel therapeutic target. Another approach is blocking angiogenesis; this strategy has been reported in a number of studies. Bevacizumab is a recombinant humanised monoclonal antibody to VEGF. It has been shown to have a role in the treatment of many cancers. It has been approved by the US FDA for advance cervical cancer. This has opened the door for future studies on anti-angiogenesis as the backbone of chemotherapy via targeting additional pathways. But patients who were treated with concurrent bevacizumab and chemotherapy developed fatigue, anorexia, hypertension, hyperglycaemia, hypomagnesaemia, headache, weight loss, and urinary tract infection. A more serious side effect, gastrointestinal perforation, was reported in 3.2 per cent of the patients treated with the drug.

The focus of this review is on potential anti-oncoproteins for therapeutic intervention as part of the new drug model. As explained earlier, interfering with oncoprotein-induced p53-degradation in order to restore the function of p53 is a promising therapeutic option for cervical cancer. A study showed that as a consequence of HPV 16 E6-induced p53 inactivation, two kinases, serum- and glucocorticoid-regulated kinase 2 (SGK2) and p21-activated kinase 3 (PAK3), played an essential role in cell proliferation and viability. When cells lost both p53 and either SGK2 or PAK3, cell death occurred. The kinase did not show similar function in non-infected human foreskin keratinocytes expressing wild-type p53. This means a novel therapeutic strategy may be developed based on synthetic lethal interaction between loss of p53 and drugs targeting either SGK2 or PAK3 in HPV-associated cervical cancer.

The zinc domain is critical for the function of E6 oncoprotein. This can be a novel target. In a study, Beerheide et al. found that the compound 4, 4'-dithiodimorpholine selectively inhibited cell viability and induced higher levels of p53 protein in tumorigenic HPV-containing cells. This compound, the authors concluded, can potentially interfere with the biology and pathology of HPV. Dithiocarbamates (DTCs) have been reported to exhibit a broad spectrum of antitumour activities. In a study, Li et al. provided evidence that DTC1 suppressed both expression of E6 mRNA and E6 oncoprotein but had no effect on the expression of E7 mRNA and protein in HPV-18. DTC1 may therefore serve as a novel chemotherapeutic agent for the treatment of cervical cancer and potential anti-HPV candidate. DTC1 has a number of abilities. It could inhibit proliferation and induce apoptosis in HeLa cells by activating caspase-3,-6, and -9. In addition, it can decrease the levels of Bcl-2 and Bcl-XL.

Furthermore, it induced HeLa cell apoptosis through a p53-dependent pathway. Duplex formation between the branch point-binding region (BBR) of U2 snRNA and the branch point sequence (BPS) in the intron is essential for splicing. Several lines of evidence indicate that splicing is functionally coupled to transcription. Both BBR and BPS interact with the U2 small nuclear ribonucleoprotein (snRNP)-associated complex. This interaction is targeted by the antitumour drug E7107, as reported by Folco et al. E7107 blocks spliceosome assembly by inhibiting the binding of U2 snRNP to pre-mRNA. The compound does not have any effect on U2 snRNP integrity, but it abolishes ATP-dependent conformational change in U2 snRNP that exposes BBR. More study is needed to understand the effect of E7 on U2 RNP, which will contribute to the establishment of the therapeutic potential of E7107, which will lead to the development of second-generation derivatives.

A number of plant compounds have also been described. Wogonin is extracted from *Scutellaria baicalensis* and is known as a benzodiazepine receptor ligand with anxiolytic effect. Studies have shown that it modulates angiogenesis, proliferation, invasion, and tumour progression in various cancer tissues. A study by Kim et al. showed that wogonin was cytotoxic to HPV-16-positive cervical cancer and that it induced apoptosis by suppressing the expression of E6 and E7 viral oncogenes. It was found that it also modulated the mitochondrial membrane potential and the expression of pro- and anti-apoptotic factors such as Bax and Bcl-2.

Finally, Luteolin is a flavonoid extracted from a number of plants with recognised anticancer, anti-inflammatory, and anti-oxidative properties; it inhibits angiogenic process and modulates multidrug resistance. A study by Ham et al. showed that Luteolin had a significant dose-dependent cytotoxic effect, only in HPV-positive cells in comparison with HPV-negative cervical cancer cells. Expression levels of HPV oncogenes E6 and E7 were suppressed, while those of related factors pRb and p53 were recovered, and E2F5 was increased by Luteolin treatment. Furthermore, Luteolin enhanced the expression of death receptors and death receptors downstream factors such as Fas/FasL, DR5/TRAIL and FADD in HeLa cells, and activated caspase cascades. Luteolin also induced mitochondria membrane potential collapse and cytochrome release and inhibited Scl-2 and Bcl-xL expression. The authors concluded that Luteolin exerts anticarcinogenic ability through inhibition of E6 and E7 expression and apoptosis by activating intrinsic and extrinsic pathways.

A variety of promising techniques are now available to aid in the search for new compounds for the development of novel anti-oncoprotein drugs for the treatment of HPV-associated cancers. Two promising methodologies will be dealt with in chapters 14 and 15. The importance of high-throughput screening (HTS) will be discussed briefly here. HTS is a specific development in laboratory automation to collect a large amount of experimental data in a relatively short time to test hundreds and thousands of compounds per day. It has been shown that an HTS process can allow up to 1,000 times faster screening (100 million reactions in ten hours) at 1-millionth a cost than conventional techniques using drop-based micro fluidics. Through this approach, one can rapidly identify active compounds, antibodies, or genes that modulate a particular biomolecular pathway. This provides starting point for drug design and understanding the interaction and role of a particular gene or pathway in pathogen-host interaction. With oncoviral infections, HTS screening could lead to the design of small molecules that can directly inhibit the oncogenic proteins that are mutated or overexpressed in specific tumour cell types. By targeting specific molecular defects found within the tumour cells, HTS may ultimately yield therapies tailored to each tumour's specific makeup.

This methodology resulted in the development of Gleevec (Matinib), an inhibitor of the breakpoint cluster region-abelsen kinase (BCR-ABL) oncoprotein found in Philadelphia chromosome-positive chronic myelogenous leukaemia, and Herceptin (trastuzumab), a monoclonal antibody targeted against oncoproteins found in metastatic breast cancers. A strategy that can be adopted in HTS involves searching for genotype-selective antitumour agents that become cidal to tumour cells only in the presence of specific oncoproteins, such as E6, or in the absence of specific tumour suppressors. Such compounds might target oncoproteins directly, or they might target other critical proteins involved in the oncoprotein-associated signal network.

The following compounds have been reported to possess synthetic cidal abilities: 1. Rapamycin analog CCI-779 in myeloma cells lacking PTEN. 2. Gleevac in BCR-ABL-transformed cells. 3. Benzoquinazolines and thiazoloinidazole for influenza viruses. Stockwell et al. used an assay format they termed cytoblot to identify a number of less-characterised compounds. They showed that cytoblots can be used for high-throughput screening of small molecules in cell-based assay. A number of HTS screening strategies are available. These include cell-based assay, cytopathic effect (CPE)-based assay, and ATP/luminescence assay. HTS is used to identify lead compounds for a wide range of therapeutic areas. Typically, a single data read is used to determine whether the test compound interacts with the target of interest. The result of HTS activity leads to the compound being grouped into a hit, if the target shows activation above a specified threshold, or a non-hit, if it falls below the threshold. The assay threshold typically depends on simple signal-to noise activities. Cell-based assay HTS screening can also be used for absorption, distribution, metabolism, elimination, and toxicity (ADMET) in the early stage of drug discovery. The CPE-based antiviral HTS assay involves cell-based screening for potential antiviral compounds that measures the CPE induced by the viral infection in a cell to measure cell viability. With luminescence assay, a luciferase gene is engineered into a Replicon or full-length viral genome to monitor viral replication. Potential inhibitors could be identified through suppression of luciferase signals upon compound incubation. With the availability of very promising screening assay, more novel ideas should be developed to facilitate the development of the magic bullet for HPV-associated carcinomas.

7.5   HPV Vaccines

So far, two prophylactic HPV vaccines are in use: Gardasil, a quadrivalent HPV 16/18/6/11 virus-like vaccine developed by Merck and Co. (NJ, USA), and Cervarix, a bivalent HPV 16/18 particle from GSK plc (London). They have been found to be efficient in preventing HR HPV infections and minimising the consequences of HPV-associated diseases. But these vaccines are only effective in girls from 11 to 12 years as recommended by the FDA, without history of past history of HPV infection and not shown therapeutic effect against current infection or associated lesions. From these, it can be argue that a large population will remain at risk of HPV infection. This makes the development of therapeutic vaccine against HR-HPV a global health priority, especially of HPV-16, which is responsible for 46 to 65 per cent of squamous cell carcinomas worldwide. In addition, HPV-58, which is the third most common virus in cervical cancer in Asia, needs to be a point of focus for therapeutic vaccine.

This section will review some of the novel ideas utilised for the development of an HPV therapeutic vaccine. A review of studies provides an overview of the efficacy and clinical effectiveness of a bivalent (HPV 16, 18), quadrivalent (HPV 6, 11, 16, 18), and 9v HPV (HPV 6, 11, 16, 18, 31, 33, 45, 52, 58) vaccines; all the three vaccines showed high efficacy in the prevention of vaccines-specific HPV-type infection and associate high-grade cervical dysplasia in HPV-naïve women. An early clinical effectiveness data for the bivalent and quadrivalent vaccine showed reduced rates of HPV 16 and 18 prevalence in vaccinated cohorts. In addition, the bivalent vaccine exhibited cross-protection to non-vaccine HPV types.

However, there is no clinical effectiveness data for 9v HPV vaccine. Apo Immune has developed a novel HPV vaccine, ApoVax104-HPV, which comprises chimeric molecules containing extracellular domain of costimulatory 4-1 BBL fused C-terminus to core streptavidin (ApoVax104) and biotinylated HPV-16 E7 oncoprotein as a TAA conjugated to the chimeric protein via biotin/streptavidin interaction. Phase I studies showed that ApoVax104 component of the vaccine targeted conjugated antigen into DCs constitutively, expressing the 4-1BB receptor, which activated the DCs for antigen uptake and presentation, which led to initiation of adaptive immunity. It also directly worked on CD4+ and CD8+ T effector (Teff) cells to augment the adaptive immunity, and most importantly, it overcame the suppressive function of CD4+CD25+FoxP3+T regulatory (Treg) cells. The pleotropic effects of 4-1BL on innate, adaptive, and regulatory immunity gives it some advantage over novel vaccines which are being developed. In a study, Wang et al. modified HPV58 E6 and E7 oncogenes to eliminate their oncogenic potential and then constructed a recombinant DNA vaccine that co-expressed the sig-HPV58ME6E7-Fc-GPI fusion antigen plus Granulocyte-Macrophage colony stimulating factor (GM-CSF) and a B7.1 serving as molecular adjuvant (PVAXI-HPV58mE6E7FcGB) for the treatment of HPV58-positive cancer.

Mutant mice were challenged by HPV58 E6E7-expressing B/6-HPV58 E6E7 cells. This was followed by immunisation with the test vaccine on days 7, 14, and 21 after the tumour challenge. They reported that the test vaccine elicited varying levels of IFN-1sgdB T-cell immune response. They therefore concluded that the test vaccine efficiently generated cellular immunity and antitumour efficacy in the immunised mice.

These novel strategies are all still in the hypothetical phase. Large populations studies are needed to assess these vaccines. However, HPV vaccines carry unique challenges. Three issues are highly essential: acceptability, dose adherence, and knowledge of healthcare providers. Vaccine acceptability is a crucial factor in vaccine uptake. Belief and willingness to be vaccinated are crucial factors. A study by Hoque and Van Hal aimed at investigating the acceptability of HPV vaccine among educated people in South Africa found that among 146 MBA students, the majority of them (74 per cent) had heard of cervical cancer, but only 26.2 per cent had heard of HPV. After reading some information on cervical cancer and HPV, the intention to have their daughter vaccinated increased from 88 per cent to 97.2 per cent. This can be attributed to fears of the disease. Those who did not want their daughters vaccinated attributed their decision to the safety of the vaccine.

Gilbert et al. reported that among 296 heterosexual men and 312 gay and bisexual men in the US, more gay and bisexual men were willing to receive HPV vaccine than heterosexual men. Gay and bisexual men reported greater awareness of HPV vaccine, perceived worry of HPV-associated diseases, perceived effectiveness of HPV vaccine, and anticipated regret if they declined vaccination and later developed HPV-associated diseases than heterosexual men. This means novel intervention strategies for the heterosexual group may be needed.

A systematic studies of fourteen unique studies representing ten sub-Saharan Africa countries reported that acceptability of HPV vaccine for females was high, but vaccine-related awareness and knowledge were low. In Ghana, it was found that of 204 women aged 18–65 who took part in a survey, 94 per cent were willing to vaccinate themselves or allow their daughters to be vaccinated. Side effects and safety of needles were some of the main concerns. In the north of Nigeria, which has been plagued by rejection of polio vaccine, Iliyasu et al. reported that among 375 female university students, 74 per cent were willing to accept HPV vaccination. Another study in Nigeria showed that 70 per cent of 201 mothers were willing to accept vaccination for their children. The 30 per cent stated sexual promiscuity as the reason for refusing HPV vaccination.

Dose adherence is also essential. Without adhering to the vaccination schedule, the success of any program will be thin. In the US, HPV vaccine has been recommended for use in girls and young women since 2007, but HPV vaccine uptake is low. A study by Liu et al. reported that among 378,484 females aged 9–26 years, only 29.4 per cent completed HPV vaccination. Another study by Hirth et al. involving 271,976 females in the US found that those aged 13 years to 18 years, 19 years to 26 years, and ≥ 27 years were less likely than those 9 years to 12 years to complete their HPV vaccine schedule. Even among men, adherence is a problem. Hirth et al. found that among 514 males who initiated HPV vaccination between 2006 and May of 2009, only 21 per cent completed all three vaccine doses within twelve months, and completion decreased over time. More of the studies reviewed showed high acceptance rates with few rejections. Reasons for refusing HPV vaccines included safety concerns, doubts about its efficacy, and religious beliefs. It is therefore imperative that HPV campaigns should address gaps in knowledge regarding HPV, genital warts, and cervical cancer. It should attend to concerns about vaccine safety and efficacy. In addition, strategies should be developed to address the issue of social stigma surrounding HPV vaccine as well as promoting families and partners support for women who decide to be vaccinated. In the words of Heidi Larson, "The world must accept that HPV vaccine is safe."

REFERENCES

ACT Against cancer: NCI announces potential HPV-related biomarker and spreader of clinical trials on www.uacancer.com/home/category/hpv 6/18/2013 (Accessed on 10th December 2015).

Agrestia JJ, et al (2010), Ultrahigh-throughput screening in drop-based micro fluidics for directed evolution. PNAS USA 107: 4004–4009.

Alonso LG, et al (2004), The HPV 16 E7 oncoprotein self assemble into defined spherical oligomer. Biochemistry 43: 3310–3317.

An WF, Tolliday N (2010), Cell-based assays for high-throughput screening. Mol Biotechnology 45: 180–186.

Arbeit JM, et al (1996), Chronic estrogen-induced cervical and vaginal squamous carcinogenesis in human papillomavirus transgenic mice. PNAS USA: 93 2930–2935.

Arbyn M, et al (2012), Evidence regarding human papillomavirus testing in secondary prevention of cervical cancer. Vaccine 30: F88–F99.

Arbyn M et al (2013), The APTIMA HPV assay versus the Hybrid capture test in triage women with ASC-US or LSIL cervical cytology: A meta-analysis of the diagnostic accuracy. Int J of Cancer 132: 101–108.

Awakumv N, Torchia J, Mymryk JS (2003), Interaction of the HPV E7 protein with the pCAF acetyltransferase. Oncogene 22:3833–3841.

Baker CC, et al (1987), Structural and transcriptional analysis of human papillomavirus type 16 sequences in cervical carcinoma cell lines. J Virol 61: 962–971.

Baldwin A, et al (2010), Kinase requirement in human cells. V. synthetic lethal interaction between p53 and the protein kinase SGK2 and PAK3. PNAS USA 107: 12463–12468.

Bao YP, et al., ACCPAB Members (2003), Human papillomavirus type distribution in women from Asia: Meta-analysis. Int J Gynecol Cancer 18: 71–79.

Beeheide W, Bernard H-U, et al (1999), Potential drugs against cervical cancer: Zinc-ejecting inhibitors of the human papillomavirus type 16 E6 oncoprotein. JNCI 91:1211–1220.

Bentley D (1999), Coupling RNA polymerase II transcription with pre-mRNA processing. Curr Opin Cell Biol 11: 347–351.

Bentley D (2002), The mRNA assembly line: Transcription and processing machines in the same factory. Curr Opin Cell Biol 14: 336–342.

Boucher J, et al (2009), Evaluation of p16INK4a minichromosom maintenance protein 2, DNA topoisomerase IIα, ProEx C, and p14INK4a/ProEx in cervical squamous intraepithelial lesions. Human Pathology 40:904–905.

Branca M, et al (2006), Relationship of up-regulation of 67-kd laminin receptor of cervical intraepithelial neoplasia and to high-risk HPV types and prognosis in cervical cancer. Acta Cytol 50: 6–15.

Branca M, et al (2006), Downregulation of E-cadherin is closely associated with progression of cervical intraepithelial neoplasia (CIN) or disease outcome in cervical cancer. Eur J Gynaecol Oral 27: 215–223.

Bravo IG, Alonso A (2004), Mucosal human papillomaviruses encode four different E5 proteins whose chemistry and phylogeny correlates with malignant or benign growth. J Virol 78: 13613–13626.

Brehma A, et al (1999), The E7 oncoprotein associates with Miz and histone deacytylase acitivity to promote cell growth. EMBO J 18: 2449–2458.

Capdeville R, et al (2002), Glivec (STI571, IMATINIB), a rationally developed, targeted anticancer drug. Nat Rev Drug Discove 1:493–502.

Castle RE, et al (2010), Relationship of atypical glandular cell cytology, age, and human papillomavirus detection to cervical and endometrial cancer risks. Obs & Gynecol 115:243–248.

Cheng S, Schmidt-Grimminger DC, et al (1995), Differentiation-dependent up-regulation of the human papillomavirus E7 gene reactivates cellular DNA replication in suprabasal differentiation keratinocytes. Gene Dev 9: 2335–2349.

Clements A, et al (2000), Oligomerization properties of the viral oncoprotein adenovirus E1A and human papillomavirus E7 and their complexes with the retinoblastoma protein. Biochemistry 26: 289–293.

Clemens KE, et al 1985), Dimerization of the human papillomavirus oncoprotein in vivo. Virology 214: 289–293.

Clifford GM, Smith JS, et al (2003), Human papillomavirus types in invasive cervical cancer worldwide: A Meta analysis. Br J Cancer 88: 63–73.

Coleman MA, et al (2011), HPV vaccine acceptability in Ghana West Africa. Vaccine 29: 3945–3950.

Cunningham MI, Davison C, Aronson KJ (2014), HPV vaccine acceptability in Africa: A systematic review. Science 69: 274–278.

Dahiya A, et al (2000), Role of the LXCXE binding site in Rb function. Mol Cell Biol 20: 6799–6805.

Das R, Dufu K, et al (2006), Functional coupling of RNAR II transcription to spliceosome assemble. Genes & Dev 20: 1100–1109.

Depuydt CE, et al (2011), BD-Pro ExC as adjunct molecular marker for improved detection of CIN+ after primary screening. Cancer Epidemiology & Prevention 20: 628–637.

Deunsing S, Munger K (2004), Mechanism of genomic instability in human cancers: Insight from studies with human papillomavirus oncoproteins. Int J Cancer 109:157–162.

de Villiers EM, et al (2004), Classification of papillomavirus. Virology 324:17–27.

Druker BJ, et al (1996), Effects of a selective inhibitor of the Abl tyrosine kinase on the growth of Bcr- Abl positive cells. Nat Medicine 2:561–566.

Dyson N, et al (1989), The human papillomavirus -16 E7 oncoprotein is able to bind to the retinoblastoma gene product. Science 243: 934–937.

Eichten A, et al (2004), Molecular pathways executing the "trophic sentinel" response in HPV-16 E7 expressing normal human diphoid fibroblasts upon growth factor deprivation. Virology 319: 81–93.

Ezeanochie MC, Olagbuji BN (2014), Human papillomavirus: Determinant of acceptability by mothers for adolescents in Nigeria. Afr J Reprod Health 18: 154–158.

Fisher CM, Schefter T (2015), Profile of bevacizumab and its potential in the treatment of cervical cancer. Onco Target 8: 3425–3431.

Folcon EG, Coil KE, Reed R (2011), The anti-tumor drug E7107 reveals an essential role for SF3b in remodeling U2 snRNP to expose the branch point-binding region. Genes & Dev 25: 440–444.

Francis DA, et al (2000), Repression of the integrated papillomavirus E6/E7 promoter is required for growth suppression of cervical cancer cells. J virol 74: 2671–2686.

Gadduci H, et al (2013), Tissue biomarkers as prognostic variables of cervical cancer. Crit Rev in Oncol & Hematol 86: 104–129.

Gage JR, et al (1990), The E7 protein of the non oncogenic human papillomavirus type 6b (HPV-6b) and of the oncogenic HPV-16 differ in retinoblastoma protein binding and other properties. J virol 64:723–730.

Garrett TO, Duerksen-Hughes PJ (2006), Modulation of apoptosis by human papillomavirus (HPV) oncoproteins. Arch Virol 151: 2321–2335.

Gilbert P, et al (2010), HPV vaccine acceptability in heterosexual, gay, and bisexual men. Am J Men Health 5: 4297–435.

Gocze K, et al (2015), MicroRNA expression in HPV-induced cervical dysplasia and cancer. Anticancer Res 35: 523–530.

Gong EY, et al (2013), Development of a high-throughput antiviral screening assay for screening inhibitors of Chikungunya virus and generation of drug-resistant mutation in cultured cells. Methods Mol Biol 1030: 429–438.

Goodwin EC, DiMaio D (2000), Repression of human papillomavirus oncogenes in HeLa cervical carcinoma cell carries the orderly reactivation of dormant tumor suppression pathways. PNAS USA 97: 12513–12518.

Griesser H, et al (2009), HPV vaccine protein L1 predicts disease outcome of high-risk HPV+ early squamous dysplastic lesions. The American J of Clinical Pathol 132:840–848.

Ham S, K Ki H, et al (2014), Luteolin induces intrinsic apoptosis via inhibition of E6/E7 oncogenes and activation of extrinsic and intrinsic pathways in HPV-18-associated cells. Oncology Rep 31:2683–2691.

Harper DM, et al., GlaxoSmithKline HPV Vaccine Study Group (2004), Efficacy of a bivalent L1 virus-like particle vaccine in prevention of infection with human papillomavirus 16 and 18 in young women: a randomized controlled trial. Lancet 364:1757–1765.

Hawley-Nelson P, et al (1989), The HPV E6 and E7 proteins cooperate to immortalize human foreskin keratinocytes. EMBO J 8: 3905–3910.

Hirose Y, Manley JL (2000), RNA polymerase II and the integration of nuclear events. Genes & Dev 14: 1415–1429.

Hirth JM, et al (2012), Completion of the human papillomavirus vaccine series among insured females between 2006 and 2009. Cancer 118: 5623–5629.

Hirth JM, et al (2013), Completion of the human papillomavirus (HPV) vaccine series among males with private insurance between 2006 and 2009. Vaccine 31: 1138–1140.

Hoque ME, Van Hal G (2014), Acceptability of human papillomavirus vaccine: A survey among master of Business Administration students in KwaZulu Natal, South Africa. Biomed Res Int Article ID: 257807.

Huang Z, Peng S, et al (2015), Combination of proteasome and HDAC inhibitor enhances HPV 16 E-specific CD8 + T cell immune response and antitumor effect in a preclinical cervical cancer model. J Biomed Sci 22:7.

Hwang SW, et al (2001), Human papillomavirus type 16 E7 binds to E2FI and activates E2FI-driven transcription in a retinoblastoma protein-independent manner. J BIol Chem 227: 2923–2930.

Iliya Z, et al (2010), Cervical cancer risk predictor of human papillomavirus acceptance among female university students in north Nigeria. J Obstet Gynaecol 30: 857–862.

Jiang P, Yue Y (2014), Human papillomavirus oncoproteins and apoptosis (Review). Exp Ther Med 7:3–7.

Jeon S, et al (1995), Integration of human papillomavirus type 16 into the human genome correlates with a selective growth advantage of cells. J virol 69: 2989–2997.

Kadaja M, et al (2009), Mechanisms of genomic instability in cells infected with the high-risk human papillomavirus. PLos Pathog 5: e10003907.

Kaspersen MD, Larsen PB, et al (2011), Identification of multiple HPV types in spermatozoa from human sperm donors. PLoS One 6: e18095.

Katich SC, et al (2001), Regulation of cdc 25A gene by the human papillomavirus 16 E7 oncogene. Oncogene 20: 543–530.

Kim MS, Bak Y, et al (2013), Wogonin induces apoptosis by suppressing E6 and E7 expression and activating intrinsic signaling pathway in HPV-16 cervical cancer cells. Cell Biol Toxicol 29:259–272.

Konarska MM, Query CC (2005), Insight into the mechanisms of splicing: More lessons from the ribosome. Genes Dev 19: 2255–2260.

Koskimaa HM, et al (2010), Molecular markers implicating early malignant events in cervical carcinogenesis. Cancer Epidemiology Biomarkers & Prevention 19: 2003–2012.

Larson H (2015), The world must accept that the HPV vaccine is safe. Nature 3:528.

Li Y, et al (2010), High risk human papillomavirus reduces the expression of microRNAs-218 in women with cervical intraepithelial neoplasia. J Int Med Res 38: 1730–1736.

Li Y, et al (2015), A novel dithiocarbamate derivative induces cell apoptosis through p53-dependent intrinsic pathway and suppresses the expression of the E6 oncogene in human papillomavirus 18 in HeLa cells. Apoptosis 20:787–795.

Lin Z, Bazzaro M, et al (2009), Combination of proteasome and HDAC inhibitor for uterine cervical cancer treatment. Clin Cancer Res 15: 570–577.

Liu G, et al (2015), HPV vaccine campaign and dose adherence among commercially insured females aged 9 through 26 years in the US. Papillomavirus Res 2:1–8.

Liu X, et al (2006), Structure of the human papillomavirus E7 oncoprotein: Self assemble into defined spherical oligomers. Biochemistry 43:3310–3317.

Longswork MS, Laimins LA (2004), Pathogenesis of human papillomaviruses in differentiating epithelial cells. Microbiol Mol Biol Rev 68: 362–372.

Luckett R, Feldman S (2015), Impact of 2-, 4-, and 9-valent HPV vaccine on morbidity and mortality from cervical cancer. Human Vaccines & Immunotherapeutic DOI: 10.1080/21645515.2015.1108500.

Luhn P, Wentzensen N (2013), HPV-based tests for cervical cancer screening and management of cervical disease. Current Obs & Gynecol Reports 2:677–700.

Lui H, et al (2012), Genomic amplification of the human telomerase gene (hTERC) associated with human papillomavirus is related to the progression of uterine cervical dysplasia to invasive cancer. Diagnostic Pathol 7:147

Maddry JA, Chen X, et al (2011), Discovery of novel benzoquinazoline and thiazoloimidazoles inhibitors of influenza H5N1 viruses, from a cell-based high-throughput screen. J Biomol Screen 16: 73–81.

Majno G, Joris I (1995), Apoptosis, oncosis, and necrosis. Am J Pathol 146: 3–15.

Malta M, et al (2015), Let-7c is a candidate biomarker for cervical cancer intraepithelial lesions: A pilot study. Mol Diagnosis & therapy 19:191–196.

McLanghlin-Drubin ME, Munger K (2009), The human papillomavirus E7 oncoprotein. Virology 384: 335–344.

McQueen F, Duvall E (2006), Using a quality control approach to define an adequately cellular liquid-based cervical cytology specimen. Cytopathology 17: 168–171.

Meijer M, et al (2009), Validation of high-risk HPV testing for primary cervical screening. J of Clinical Virology 46 (Suppl 3), S1–S4.

Mokbel K, Hassanally P (2001), From HER2 to Herceptin. Curr Med Res Opin 17: 51–59.

Moody CA, Laimins LA (2010), Human papillomavirus oncoprotein pathway to transformation. Nature Rev 10:530–560.

Mishra A, Verma M (2010), Cancer biomarkers: Are we ready for the prime time? Cancers 2: 190–208.

Munger DJ, et al (1988), Human papillomavirus type 16 alters human epithelial cell differentiation in vitro. PNAS USA 85: 7169–7173.

NCI : http://www.cancer.gov/publications/dictionaries/cancer-terms/?CdrID=45618 (Accessed on December 14, 2015)

Nichols AC, Palma DA, et al (2013), High frequency of activating PIK3CA mutation in human papillomavirus-positive orpharyngeal cancer. JAMA Otolaryngol Head Neck Surg 139: 617–622.

Nyugen CL, Munger K (2008), Direct association of the HPV 16 E7 oncoprotein with cyclin A/CDK2 and cyclin E/CDK2 complexes. Virology 380: 21–25.

Nyugen DX, et al (2002), Human papillomavirus type 16 E7 maintained elevated levels of the cdc 25A tyrosine phosphatase during deregulation of cell cycle arrest. J Virol 76: 619–632.

Patel DA, Patel AC, et al (2012), High-throughput for small molecules enhancers of the interferon signaling pathway to drive next-generation antiviral drug discovery. PLoS One 7: e36594.

Patel DA, Patel AC, et al (2014), High-throughput screening normalized to biological response: Application to antiviral drug discovery. J Biomed Screen 19.10.117711087957113496848.

Patrick DR, et al (1994), Identification of a novel retinoblastoma gene product binding site in human papillomavirus type 16 E7 protein. J Biol Chem 269: 6842–6850.

Peng S, et al (2004), Development of a DNA vaccine targeting human papillomavirus type 16 oncoproteins E6. J Virol 78: 8468–8476.

Phelps WC, et al (1988), The human papillomavirus type 16 E7 gene encodes transactivation and transformation function similar to those of adenovirus E1A. Cell 53: 539–547.

Pisani P, Bray F, Parkin DM (2002), Estimates of the world-wide prevalence of cancer for 25 sites in the adult population. Int J Cancer 97:72–81.

Puig-Basagoiti F, Deas TS, et al (2005), High-throughput assays using a luciferase expressing Replicon, virus-like particles, and full length virus for West Nile virus drug discovery. Antomicrob Agents Chemother 49: 4980–4988.

Rasmussen L, Maddox C, et al (2011), A high-throughput screening strategy to overcome virus instability. Assay Drug Dev Technol 9: 184–190.

Rauber D, et al (2008), Prognostic significance of the detection of human papillomavirus L1 protein in smears of mild to moderate cervical intraepithelial lesions. Eur J Obs & Gynecol 140: 258–260

Ressing ME, Sette A, et al (1995), Human CTL epitopes encoded by human papillomavirus type 16 E6 and E7 identified through in vivo and in vitro immunogenicity studies of HLA-A*0201-binding peptides. J Immunol 154:5934–5943.

Ribeiro J, et al (2015), miR-34a and miR-125b expression in HPV infection and cervical cancer development. Biomed Res Int 2015:304584.

Ronco G, et al (2010), Efficacy of human papillomavirus testing for the detection of invasive cervical cancer and cervical intraepithelial neoplasia: A randomized controlled study. The Lancet Oncol 11: 249–257.

Saha B, et al (2007), Telomerase and markers of cellular proliferation are associated with the progression of cervical intraepithelial neoplasia lesions. Int J Gynecol 26: 214–222.

SBIR-STRR: Apovax104- HPV as a novel vaccine for cervical cancer on https://www.sbir.gov/sbirsearch/detail/91754 (Accessed on 13th December,2015).

Schiffman M, et al (2005), The carcinogenicity of human papillomaviruses types reflects viral evolution. Virology 357: 76–84.

Schiker JT, et al (2008), An update of prophylactic human papillomavirus L1 virus-like particle vaccine clinical trials results. Vaccine 26: 53–61.

Schlecht NF, Kulagu S, et al (2001), Persistent human papillomavirus as a predictor of cervical cancer intraepithelial neoplasm. HAMA 286: 3406–3114.

Schwarz E, et al (1985), Structure and transcription of human papillomavirus sequences in cervical carcinoma cells. Nature 314:111–114.

Shi Y, et al (2002), Enhanced sensitivity of multiple myeloma cells containing PTEN mutations to CCGI-779. Cancer Res 62: 5027–5034.

Smith DJ, et al (2008), "Nought may endure but mutability": Spliceosome dynamics and the regulation of splicing. Mol Cell 30: 657–666.

Smith PP, et al (1992), Viral integration and fragile sites in human papillomavirus-immortalized human keratinocytes cell lines. Genes Chromosome Cancer 5:152–157.

Stockwell BR, et al (1999), High-throughput screening of small molecules in miniaturized mammalian cell-based assays involving post-translational modifications. Chem Biol 6: 71–83.

Stockwell BR (2004), The biological magic behind the bullets. Nat Biotechnology 22: 37–38.

Stoilov P, et al (2008), A high-throughput screening strategy identifies cardotomic steroids as alternative splicing modulators. PNAS USA 105: 11218–11223.

Stubenrauch F, et al (1998), Differentiation requirement for conserved E2 binding sites in the life cycle of oncogenic human papillomavirus type 31. J virol 72: 1071–1077.

Tewari KS, et al (2014), Improved survival with bevacizumab in advanced cervical cancer. NEJM 370:734–743.

The Biomarkers Consortium on www.biomarkersconsortium.org (Accessed on December 14, 2015)

Thierry F, Yaniv M (1987), The BOV1-E2 trans-acting protein can be either an activator or a repressor of the HPV-18 regulatory region. EMBO J 6:3391–3397.

Tornesello ML, et al (2011), Viral and cellular biomarkers in the diagnosis of cervical intraepithelial neoplasia and cancer. Biomed Res Int Article ID: 519619.

Van Doorslaer K, Burk RD (2012), Association between hTERT activation as HPV E6 protein and oncogenic risk. Virology 433:216–219.

Venuti A, Paolini F, et al (2011), Papillomavirus E5: The smallest oncoprotein with many functions. Mol Cancer 10:40.

Villa LL, et al (2005), Prophylactic quadrivalent human papillomavirus (types 6, 11, 16, and 18) L1 virus-like particle vaccine in young women: A randomized double-blind placebo-controlled multicenter phase II efficacy trial. Lancet Oncol 6:271–278.

Wahl MC, et al (2009), The spliceosome: Design principles of a dynamic RNP machine. Cell 136: 701–718.

Walboomen JM, et al (1999), Human papillomavirus is a necessary cause of invasive cervical cancer worldwide. J Pathol 189: 12–19.

Wang H, Yu J, Li L (2015), A DNA vaccine encoding mutate HPV 58mE6E7-Fc-GPI fusion antigen and GM-CSF and B7.1. Oncogene 8: 3062–3077.

Wieringa HW, et al (2015), Breaking the DNA damage response to improve cervical cancer treatment. Cancer Treat Rev pii: S0305–7372 [Epub ahead of print].

Xiao H, et al (2014), Expression of Yes-associated protein in cervical carcinoma epithelial lesions. Int J Gynecol Cancer 24: 1575–1582.

Xu J, et al (2012), Expressed miR-24 expression via Upregulation of target gene chk1 contributes to the progression of cervical cancer. Oncogene 32: 976–987.

Yuan CH, Filippova M, Duerksen-Hughes P (2012), Modulation of apoptotic pathways by human papillomavirus (HPV), mechanisms and implication for therapy. Viruses 4: 3831–3850.

Zang R, Li D, et al (2012), Cell-based assays in high-throughput screening discovery. Int J Biotech for Wellness industries 1: 31–51.

Zerfass-Thomas K, et al (1996), Inactivation of the cdk inhibitor p27KIP1 by the human papillomavirus type 16 E7 oncoprotein. Oncogene 13: 2323–2330.

Zimet GD, Rosenthal SL (2010), HPV vaccine among males: Issues and challenges. Gynecol Oncol 117: (2 Suppl 1): S26–31.

zur Hausen H (2002), Papillomavirus and cancer: From basic studies to clinical application. Nat Rev 2: 342–350.

Zur Hausen H (2009), Papillomavirus in the causation of human cancers: A brief historical account. Virology 384: 260–265.

CHAPTER 8

VIRUS-HOST INTERACTION IN HCV-ASSOCIATED HEPATOCELLULAR CARCINOMA AND POTENTIAL FOR HCV INHIBITORS

8.1 Introduction

Hepatitis C virus (HCV) infection was first suspected in the 1970s when most blood transfusion infections were associated with either hepatitis A or hepatitis B virus. The viral genome was identified in 1989 and named hepatitis C. HCV belongs to the Hepacivirus, *Flaviviridae* family, and has six major genotypes and more than seventy subtypes. The HCV genome encodes a single precursor polyprotein of about three thousand amino acids, which is cleaved into co- and post-translationally into functional structural and non-structural proteins by host and viral protease, including three structural proteins: the core proteins forming the viral nucelocapsid and two envelope glycoproteins: E1 and E2. HCV particles are about 55–60 nm in diameter. The E1 and E2 are type I transmembrane glycoprotein composed of up to six and eleven potential glycosylation sites, respectively, as well as forming noncovalent heterodimers. The nucleocapsid is probably made of multiple copies of core protein in complex with the viral genome and lies beneath the envelope. The nonstructural proteins are NS2, NS3, NS4A, NS4B, NS5A, and NS5B.

HCV infection is a major public health issue, with an estimated 170 million people infected globally and approximately 3 to 4 million new infections yearly. Most of the patients infected are unaware of their status and remain asymptomatic; however, the majority will progress to chronic HCV. The infection by HCV may be unresolved in about 85 per cent of the infected individuals, which represents an important cause of liver cirrhosis and hepatocellular carcinoma (HCC). HCV is the most common cause of chronic liver disease and cirrhosis in the world as well as the major cause of liver transplantation in USA, Australia, and Europe.

The chronic infection is evidenced by the presence of HCV-RNA in the blood at least six months after viral contamination. HCV infects the hepatocytes and other cells such as leukocytes and epithelial cells of different organs. However, it does not cause cytotoxicity, which suggests that both the hepatitis injury and extra-hepatic clinical manifestations caused by HCV infection are probably mediated by the immune evasion of cryoglobulinemia, and complex persistence and autoimmune recognition. This means that the pathogenesis of HCV infection involves a complex virus/host interaction, which is the focus of this chapter.

8.2 Association of HCV and HCC

HCC accounts for 85 to 90 per cent of all cases of primary liver cancer. Chronic hepatitis and cirrhosis constitute most of the tumour preneoplastic conditions in the majority of HCC. The risk of developing HCC for patients with HCV-associated cirrhosis is about 2–6 per cent per year. HCC risk increases to 17-fold in HCV-infected patients compared to HCV-negative patients. In general, HCC develops after two or more decades of HCV infection, and the increased risk is restricted to patients with cirrhosis or advanced fibrosis. It was explained earlier that multiple steps are needed for the induction of all cancers. For hepatocarcinoma, it is mandatory that genetic mutation should accumulate in the hepatocytes. In HCV infection, however, it has been found that some of these steps might be skipped in the development of HCC in the presence of the core protein. The overall effects achieved by the expression of the core protein are the induction of HCC, even in the absence of a complete set of genetic changes required for carcinogenesis.

HCC is among the leading causes of cancer-associated mortality and the third most common cause of cancer deaths in the world. In Japan, for example, unlike other Asian countries, there are high incidences of HCC caused by HCV infection, accounting for 80 to 90 per cent of all cases, while in Western countries, HCC is known to complicate cirrhosis secondary to hepatitis C in 2–6 per cent of cases per year. Currently, there is no evidence to show that HCV by itself is oncogenic; however, HCC rarely develops in non-cirrhotic HCV-infected individuals. This means a direct oncogenic effect cannot be ruled out.

However, in the pathogenesis of HCV-associated HCC, it remains controversial whether the virus plays a direct or indirect role. Studies using transgenic mouse models in which the core protein of HCV has oncogenic potential showed that HCV is directly involved in hepatocarcinogenesis, although other factors, such as continued cell death and regeneration associated with inflammation, may have a role. HCV causes HCC via an indirect pathway by causing chronic inflammation, cell death, proliferation, and cirrhosis. HCV genome has been detected in the tumour and surrounding liver tissue. Regarding association with the genotype of HCV and HCC, the incidence of genotype 1b has been shown to be markedly high among patients where it is associated with a more rapid determination of the liver histology in chronic hepatitis.

Some studies, however, have argued against this proposition. Prospective studies undertaken to establish whether inflammation with specific HCV genotype was associated with an increased risk of development of HCC in cirrhosis reported that cirrhotic patients infected with HCV ib carry significant higher risk of developing HCC than patients infected by other HCV types. There is the suggestion that the presence of HBV gene in patients with chronic HCV-associated liver injury appears to promote hepatocarcinogenesis.

8.3   HCV/Host Interaction

Just like other viruses, HCV absolutely depends on the host cell for replication. Studies of the HCV life cycle (figure 8.1) have been hampered by lack of efficient cell culture systems to generate infectious viral particles in vitro. However, using a number of model systems,   CV   envelope glycoprotein, HCV-like particles (HCV-LPs), HCV cell culture (HCVcc), and retroviral HCV pseudotypes (HCVpp), it was shown that E1 and E2 are critical for host cell entry. Studies using monoclonal or polyclonal antibodies targeting both the linear and conformational epitopes of E2 showed that there was inhibition of cellular binding of HCV-LP binding, entry of HCVpp, and infection of HCVcc, which suggest that E2 has essential role for host cell surface interaction.

There are hypervariable regions within the E2 envelope glycoprotein sequence. These amino acids differ by up to 80 per cent among the HCV genotypes and even differ among subtypes of the same genotypes. The N-terminal 27 residues of E2 show high degrees of variation, and this portion of the sequence is termed hypervariable regions 1 (HVR-1). This region plays a crucial role in HCV interaction with the host cell.

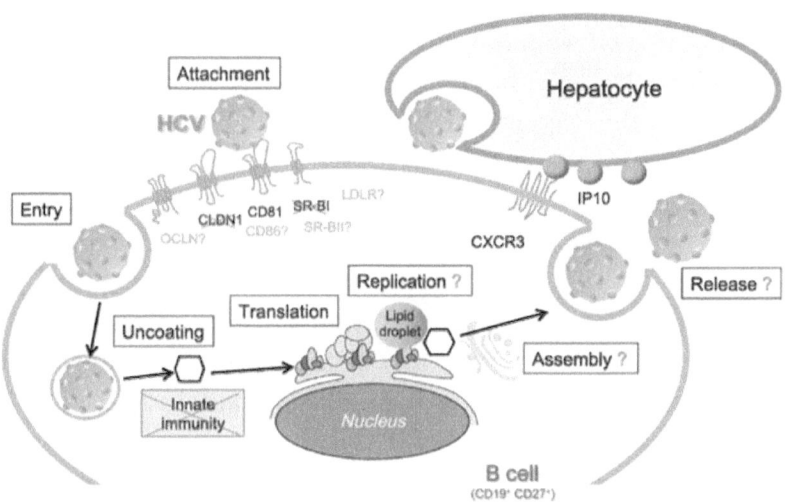

Figure 8.1: HCV Life cycle

A number of studies have demonstrated the role of HVR-1 in HCV infectivity. Studies have shown that antibodies targeting regions within HVR-1 inhibit cellular recombinant E2, HCV-LP binding, and HCVpp entry into target cells. The exact role of E1 remains unknown, but it has been proposed that E1 may directly interact with cell surface molecules or contribute to proper folding and processing of E2. It has been shown that antibodies targeting the N-terminal region of E1 inhibit HCV-LP binding and HCV infection of B-cell-derived cell lines, which suggests that E1 cell surface interaction may contribute to viral binding and entry. Furthermore, it has been shown that HCV envelope E1 and E2 induce fusion between the viral envelope and host cell membrane. HCV entry into hepatocytes is a multistep process which involves the utilisation of multiple host molecules, such as low density lipoprotein receptor (LDLR), tetraspanin CD81, scavenger receptor class B type I (SR-BI), tight junction (TJ) proteins, Claudi-1 (CLDNI), the cholesterol uptake receptor Niemann-Pick C1-like1 (NPC1L1), and occludin (OCLN).

In addition, receptor tyrosine kinase epidermal growth factor receptor and ephirin receptor AZ have been identified as HCV entry cofactors. Martin and Uprichard also identified transferring receptor 1 as a HCV entry factor. Several studies analysing interaction of HCV proteins with host cellular proteins utilising proteomic analysis showed that at least 420 host proteins interact with Jak/STAT, insulin, TGF β, and focal adhesion molecule pathways, with the majority of these interacting with core proteins, NS3, and NS5A. It is interest to note that NS5A interacts with many proteins implicated in signal transduction, cell growth and death, and cancer. Data shows that viral proteins play a central role in regulating metabolic processes, cell-to-cell adhesion, and cytoskeletal organisation, leading to HCV pathogenesis. Some of these interactions play an essential role in viral replication, eliciting strong cellular and humoral immune response, developing strategies to evade immune recognition, and in the induction of hepatitis leading to cirrhosis and HCC.

HCV has developed different mechanisms of hijacking protein-1 and its associated proteins, DEAD box helicases DDX3 and DDX6, as well as heteronuclear protein A1, which are all associated with viral replication, assembly, and virus egress. The viral protein's core, NS2, NS3/4A, and NS5A, have been shown to be interacting with key oncoproteins and contributing to the development of HCC in HCV-infected patients, although understanding the mechanism of such interaction remains elusive.

After the establishment of persistence infection, some of the viral proteins interact with the host cellular proteins, resulting in a change of their properties and functions. Due to this, the cells and extracellular matrix components change over a period of time, leading to the remodeling of the tissue and loss of control as well as regulation of cell proliferation. Marked induction of ROS in infected cells leads to oxidative stress and suppression of host immunity by viral proteins, as reported by some studies. Also, HCV proteins interact with the cellular molecules regulatory gene YB-1, p53, and cyclin D1, resulting in the induction of liver cancer. HCV constantly evolves new variants during persistent infection, with the host being constantly subjected to episodes of hepatitis. These variants evade immune recognition, alter interactions with the host cell proteins, and induce chromosomal abnormalities, which result in liver cancer. Asialoglycoprotein receptor has also been suggested to play a role in HCV entry based on interaction between this protein and baculovirus-expressed HCV structural proteins. However, the importance of this data in the context of functional HCV entry has not been elucidated.

8.4   Entry Inhibitors

As outlined earlier, HCV is a global public health issue associated with chronic hepatitis, liver cirrhosis, and HCC. The only approved treatment is a combination therapy with IFN-α and ribavirin, which targets cellular pathway. However, sustained virologic response is achieved IN only about half of patients treated. There is an urgent need for the development of novel drug targets against HCV. Although most of the research focuses on the development of HCV-specific antivirals such as protease and polymerase inhibitors, cellular targets could be ideal for the development of a broad spectrum of antivirals.

One strategy is silencing of cellular proteins. In a study, it was found that silencing some proteins using RNA interference (RNAi) decreased HCV infectious titers by 42-fold,which means that the cellular host proteins could be associated with HCV replication, release, or viral entry. Also, targeting the same interaction patterns as small molecule inhibitors or specific antibodies might increase antiviral efficacy. Another study showed that antibody-mediated blocking of HCV co receptor CD81 prevented infection with HCV. A study by Randell et al showed that RNAi machinery exerted a proviral effect on the HCV life cycle because silencing Dicer and components of RNA-induced silencing complex RISC (Argonaute proteins EIF2C1-4) decreased HCV infectious titer.

Another novel approach is using potential inhibitors of viral attachment. As found in HIV infection, lactin cynanovirin-N (CV-N) is an active compound and was shown to possess antiviral activity against other envelope viruses. The antiviral ability is due to interaction between CV-N and high mannose oligosaccharides in the viral envelope glycoprotein. It has been shown that the envelope of HCV is highly glycosylated, containing oligomannose glycans. Oligomannose glycans interact with CV-N leading to HCV antiviral activity by blocking HCV entry into target cells. Because HCV glycosylate sites are highly conserved, drugs that target glycans on HCV glycoprotein may not lead to the rapid development of viral escapes or resistance, as found in other carbohydrate-binding agents such as plant lectins, mAb, and other specific non-peptide antibiotics such as Predimicin A, which is an HCV infectivity inhibitor. These substances can be used efficiently against viruses that need glycosylated envelopes for entry into target cells.

Another potential approach is the utilisation of heparin-derived molecules because heparin has been shown to potently inhibit HCV E2, HCVpp, HCV-LP, and HCVcc binding to hepatoma cells. Inhibitors of EGFR (erlotinib), EphA2 (dasatinib), and NPCILI (ezetimibe) have been licensed and shown to inhibit HCV entry in vitro. In addition, a small inhibitor of SR-BI (ITX 5061) is in an advanced clinical study as an HCV entry inhibitor. We need to increase our understanding of the complex viral entry process to enable us to develop new therapeutic targets to prevent the virus from reaching its site of replication. Both the viral and host cell components involved in virus entry may serve as targets for the development of HCV entry inhibitors. However, targeting viral proteins rather than host cell proteins is more ideal because of potential adverse events resulting from interference with normal cell functions. Future anti-HCV therapies should be a combination of drugs that target distinct steps of HCV infection (figure 2).

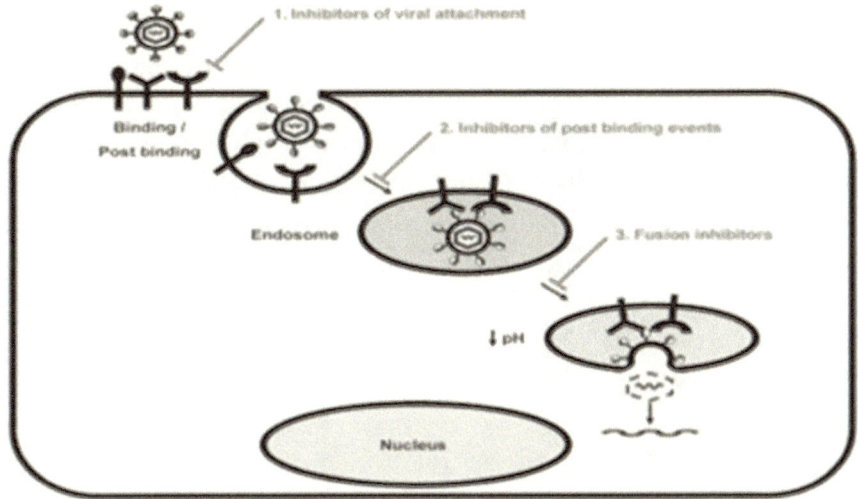

Figure 8.2: Targets of HCV infectivity inhibition

One of the major limitations of studying HCV-host interaction is the limited availability of a number of HCV strains and cell types used to characterise them. There are seven known HCV genotypes, with a nucleotide sequence diversity of about 35 per cent. But viral infection clones are available for only four genotypes: 1a and 1b, 2a, 3a, and 4a. Of these, only genotypes for 2a JFH1 and adapted genotypes 2a J6, 2b J8, and 1a H7 33 complete the viral life cycle in cell culture.

A recent advance is the development of chimeric HCV, which contains JFH1 nonstructural genes that have been fused to structural genes for all seven HCV genotypes. It allows for cross-genotyping to compare virus-host interaction with HCV core, E1, E2, p7, and NS2. For example, it has been found that infection with all seven genotype chimers can be neutralised with antibodies to CD81 and SR-BI. A number of in vitro studies have been undertaken to validate host cell factors important for HCV replication mostly in single human hepatoma cell line, Huh-7, and other sublines termed Huh-7.5, Huh-7.5.1, and Huh-7 lunet cells. There has been some progress in developing polarised cell models for HCV entry, such as Caco-2 colorectal adenocarcinoma cells that develop columnar polarity and the HepG2 hepatoma cells that develop complex hepatic polarity. Primary hepatocytes from patients can be productively infected with HCV, but there are some of the limitations, including high variability between patients and limited access to fresh hepatocytes. An ideal approach is using pluripotent stem (iPS) cells. Numerous HCV-host interaction factors have been identified despite the experimental hurdles, but new models for studying HCV-host interaction are needed to address some of the intriguing host factors, which will lead to a better understanding of this interaction.

REFERENCES

Albert A, et al (1999), Natural history of hepatitis C. J Hepatol 31: 17–24.

Baranov IN, et al (2005), Serum amyloid A binding to ClA-1 (CD36 and LIMPII analogous-1) mediates serum amyloid A protein induced activation of ERK1/2 and p38 mitogen-activated protein kinases. J Biol Chem 280: 8031–8040.

Barth H, et al (2003), Cellular binding of hepatitis C virus envelope glycoprotein E2 requires cell surface heparan sulfate. J Biol Chem 278: 41003–41012.

Barth H, et al (2005), Scavenger receptor class B type 1 and hepatitis C virus infection of primary tupaia hepatocytes. J Virol 79: 5774–5785.

Barth H, et al (2006), Hepatitis C virus: Molecular biology and clinical implications. Hepatology 44: 527–535.

Bartosch B, et al (2003), Infectious hepatitis C virus: Pseudo-particles containing functional E1 and E2 envelope protein complexes. J Exp Med 197: 633–642.

Bartosch B, et al (2003), In vitro assay for neutralizing antibody to hepatitis C virus: Evidence or broadly conserved neutralization epitopes. PNAS USA 100: 14199–14204.

Baumet TF, et al (1998), Hepatitis C virus structural protein assemble into virus like particles in insect cells. J Virol 72: 3827–3836.

Blight KJ, et al (2002), Highly permissive cell lines for subgenomic and genomic hepatitis C virus RNA replication. J Virol 76:13001–13014.

Blonski W, Raddy KR (2008), Hepatitis C virus infection and hepatocellular carcinoma. Clin Liver Dis 12: 661–674.

But DY, et al (2008), Natural history of hepatitis-related hepatocellular carcinoma. World J Gastroenterol 14: 1652–1656.

Callans N, et al (2005), Basic residues in hypervariable region 1 of hepatitis C virus envelope glycoprotein e2 contribute to virus entry. J Virol 79: 15331–15441.

Chen SL, Morgan TR (2006), The natural history of hepatitis C virus (HCV) infection. Int J Med Sci 3: 47–52.

Cormier EG, et al (2004), CD81 is an entry coreceptor for hepatitis C virus. PNAS USA 107: 7270–7274.

De Chassey B, et al (2008), Hepatitis C virus infection protein retinitis. Mol System Biology, volume 4: article 230.

Dodd RY, et al (2002), Current prevalence and incidence of infectious disease markers and estimated window-period risk in the American Red Cross donor population. Transfusion 42: 975–979.

Donato F, et al (2002), Alcohol and hepatocellular carcinoma: The effect of lifetime intake and hepatitis virus infections in men and women. Am J Epidemiol 155: 323–331.

EASL International Conference on Hepatitis C (1999) Consensus statement. European Association for the study of liver. J Hepatol 30: 956–961, Paris, 26–28, February.

El-Serag HB (2002), Hepatocellular carcinoma and hepatitis C in the United States. Hepatology 36: S74–S83.

El-Serag HB, Rudolph KL (2007), Hepatocellular carcinoma: Epidemiology and molecular carcinogenesis. Gastroenterology 132: 2557–2576.

Fujuoka S, et al (2003), Hepatitis B virus gene in liver tissue promotes hepatocellular carcinoma development in chronic hepatitis C patients. Dig Dis Sci 48: 1920–1924.

Gerber MA, et al (1992), Detection of replicative hepatitis C virus genome in hepatocellular carcinoma. Am J Pathol 141: 1271–1277.

Gottwein JM,et al (2010), Novel infectious cDNA clones of hepatitis C virus genotype 3a (strain S52) and 4a (strain ED43), genetic analyses and in vivo pathogenesis studies. J Virol 84:5277–5293.

Gottwein JM, et al (2009), Development and characterization of hepatitis C virus genotype 1–7 cell culture systems: Role of CD81 and scavenger receptor class B type I and effect of antiviral drugs. Hepatology 49:364–377.

Guntaka RV, Paola MK (2014), Interaction of hepatitis C virus protein with cellular oncoproteins in the induction of liver cancer. ISRN Virology doi: 10.1155/2014/351407.

Hoofnagle T (2004), Causes and outcomes of hepatitis C. Hepatology 36: S21–S29.

Horner SM, Gale Jr. M (2013), Regulation of hepatitis innate immunity by hepatitis C virus. Nat Medicine 19: 879–888.

Hu Y, et al (2007), Emerging host cell target for hepatitis C therapy. Drug Discovery Today 12:209–217.

Ivanov AV, et al (2013), HCV and oxidative stress in the liver. Viruses 5: 439–469.

Kamal SM (2008), Acute hepatitis C: A systematic review. Am J Gastroenterol 103: 1283–1297.

Keck ZYV, et al (2004), Human monoclonal antibody to hepatitis C virus E1 glycoprotein that blocks virus attachment and viral infectivity, J Virol 78: 7257–7263.

Kiyosama K (2002), Trend in liver cirrhosis as precancerous lesions. Hepatol Res 24: 40–45.

Koike K, et al (2008), Molecular basis for the synergy between alcohol and hepatitis C virus in hepatocarcinogenesis. J Gastroenterol Hepatol 23: S87–S91.

Koutsoudakis G, et al (2006), Characterization of the early steps of hepatitis C virus infection using luciferase reporter viruses. J Virol 80: 5308–5320.

Koutsoudakis G, et al (2007), The level of CD81 cell surface expression is a key determinant for productive entry of hepatitis C virus into host cells. J Virol 81:588–598.

Lee CM, et al (2008), Hepatitis C virus genotype 1b as a major risk factor associated with hepatocellular carcinoma in patients with cirrhosis: A seventeen-year prospective cohort study. Hepatology 46: 350–356.

Leoung TY, Leoung AS (2005), Epidemiology and carcinogenesis of hepatocellular carcinoma. HPB (oxford) 7: 5–15.

Lupberger J, et al (2011), EGFR and EphAZ are host factors for hepatitis C virus entry and possible targets for antiviral therapy. Nat Med 17: 589–595.

Machida K, et al (2009), Hepatitis C virus causes uncoupling of mitotic checkpoints and chromosomal polyploidy through the Rb pathway. J of virology 23:128590–12600.

Martin DN, Uprichard SL (2013), Identification of transferring receptor 1 as a hepatitis C virus entry factor. PNAS USA 110: 10777–10782.

Mee CJ, et al (2008), Effect of cell polarization on hepatitis C virus entry. J Virol 82:461–470.

Mee CJ, et al (2009), Polarization restricts hepatitis C virus entry into HepG2 hepatoma cells. J Virol 83:6211–6221.

Mitra AK (1999) Hepatitis C-related hepatocellular carcinoma: prevalence and the world, factors interacting, role of genotypes. Epidemiologic Review 21: 180–187.

Molina-Jimenez F, et al (2012), Matrigel-embedded 3D culture of Huh-7 cells as a Hepatocy te-like polarized system to study hepatitis C virus cycle. Virology 425:31–39.

Moradpour D, Penin F (2013), Hepatitis C virus protein: From structure to function. Current Topic in Microbiology and Immunology 369: 113–142.

Murray CL, Rice CM (2011), Turning hepatitis C into a real virus. Ann Rev of Microbiology 65: 307–321.

Op De Beeck A, et al (2005), The transmembrane domain of hepatitis C virus envelope glycoprotein E1 and E2 play a major role in heterodimerization. J Biol Chem 275: 31428–31437.

Parkin AM, et al (2000), Global cancer statistics, 2002. CA Cancer J Dis 31: 17–24.

Penin F, et al (2004), Structural biology of hepatitis C virus. Hepatology 39:5–19.

Pileri P, et al (1998), Binding of hepatitis C virus to CD81. Science 282:938–941.

Ploss A, et al (2010), Persistent hepatitis C virus infection in microscale primary human hepatocyte cultures. Proc Natl Acad Sci U S A 107:3141–3145.

Podevin P, et al (2010), Production of infectious hepatitis C virus in primary cultures of human adult hepatocytes. Gastroenterology 139:1355–1364.

Randell G, e t al (2007), Cellular cofactors affecting hepatitis C infection and replication. PNAS USA 104: 12884–12889.

Roelandt P, et al (2012), Human pluripotent stem-cell derived hepatocytes support complete replication of hepatitis C virus. J Hepatol 57:246–251.

Rosa D, et al (1996), A quantitative test to estimate neutralizing antibodies to the hepatitis C virus: Cytofluorimetric assessment of envelope glycoprotein 2 binding to target cells. PNAS USA 93: 1789–1763.

Sainz B Jr, et al (2012), Identification of Niemann-Pick C1-like I cholesterol absorption receptors as a new hepatitis C virus entry factor. Nat Med 18: 281–285.

Sangiovanni A, et al (2004), Increased survival of cirrhotic patients with a hepatocellular carcinoma detected during surveillance. Gastroenterology 126: 1005–1014.

Scarselli E, et al (2002), The human scavenger receptor class B type I is a novel candidate receptor for the hepatitis C virus. EMBO J 21: 5017–5025.

Scheel TK, et al (2008), Development of JFH1-based cell culture systems for hepatitis C virus genotype 4a and evidence for cross-genotype neutralization. Proc Natl Acad Sci U S A 105:997–1002.

Schwartz RE, et al (2012), Modeling hepatitis C virus infection using human induced pluripotent stem cells. Proc Natl Acad Sci U S A 109:2544–2548.

Selimovic D, et al (2012), Hepatitis C virus-related hepatocellular carcinoma: An insight into molecular mechanisms and therapeutic strategies. World J of Hepatology 4: 342–355.

Shenovy SR, et al (2001), Selective interaction of the human immunodeficiency virus protein cyanovirin-N with high-mannose oligosaccharide on gp 120 and other glycoproteins. J Pharmacol Exp Ther 297: 704–710.

Shulla A, Randall G (2012), Hepatitis C virus- host interaction, replication, and viral assembly. Current Opin Virol 2: 719–726.

Steinmann D, et al (2004), Inhibition of hepatitis C virus-like particle binding to target cells by antiviral antibodies in acute and chronic hepatitis C. J Virol 78: 9030–9040.

Summonds P (2004), Genetic diversity and evolution of hepatitis C virus 15 years on. J Gen Virol 85: 3173–3188.

Syder AJ, et al (2011), Small molecules scavenger receptor BI antagonists are potent HCV entry inhibitors. J Hepatol 54: 48–55

Takayama T, et al (1990), Malignant transformation of adenomatous hyperplasia to hepatocellular carcinoma. Lancet 336: 1150–1153.

Takikawa S, et al (2000), Cell fusion activity of hepatitis C virus envelope protein. J Virol 74: 5066–5074.

Tan A, et al (2008), Viral hepatocarcinogenesis: From infection to cancer. Liver Int 28:175–188.

Thirome R, et al (2001), Determinant of viral clearance and persistence during acute hepatitis C virus infection. J Exp Med 194: 1395–1406.

Triyatni M, et al (2002), Interaction of hepatitis C virus-like particle and cells: A novel system for studying viral binding and entry. J Virol 76: 9335–9344.

Tsutsumi T, et al (2003), Hepatitis C virus core protein activates ERK and p38 MAPK in cooperation with ethanol in transgenic mice. Hepatology 38: 820–828.

Wakita T, et al (2005), Production of infectious hepatitis C virus in tissue culture from a cloned viral genome. Nat Med 11:791–796.

Walker CM (2010), Adaptive immunity to the Hepatitis C virus. Advances in Virus research 78: 43–84.

Walzer N, Kulik LM (2000), Hepatocellular carcinoma: Latest development. Curr Opin Gastroenterol 24: 312–319.

Waser BM, Tellinghulsen TL (2011), Structural biology of the hepatitis C virus proteins. Drug Discovery Today: Technologies 9: 195–204.

Weiner AJ, et al (1991), Variable and hypervariable domains are found in the regions of HCV corresponding to the flavivirus envelope and NS-1 protein among the pestinium envelope glycoprotein. Virology 180: 842–848.

Wei X, et al (2003), Antibody neutralization and escape by HIV-1. Nature 422: 307–312.

Wellnitz S, et al (2002), Binding of hepatitis C virus-like particle derived from infectious clone H77C to defined human cell lines. J Virol 76: 1181–1193.

World Health Organization (1999), Hepatitis C-global prevalence (update). Weekly Epidemiologic record 74: 425.

Wu X, et al (2012), Productive hepatitis C virus infection of stem-cell derived hepatocytes reveals a critical transition to viral permissiveness during differentiation. PLoS Pathog 8:e1002617.

Zhong J, et al (2005), Robust hepatitis C virus infection in vitro. PNAS U S A 102:9294–9299.

CHAPTER 9

KAPOSI'S SARCOMA-ASSOCIATED HERPES VIRUS AND ONCOGENES

9.1 Introduction

Kaposi's sarcoma-associated herpes virus (KSHV), also referred to as human herpes virus 8 (HHV8), was first isolated from the Kaposi's sarcoma (KS) lesion of a patient with AIDS. It is a DNA virus classified as a member of the gamma 2-herpes virus subfamily. Earlier studies had confirmed that KSHV was the etiologic agent of KS. KS is a neoplasm derived from lymphatic endothelial cells infected with KSHV, made up of spindle-shaped cells and inflammatory mononuclear cells. KS is grouped into four epidemiological types: classic, endemic, iatrogenic, and AIDS-associated. KSHV is also associated with primary effusion lymphoma (PEL) and multicentric Castleman's disease (MCD)

9.1.1 Kaposi's Sarcoma (KS)

KS is a vascular tumour which is made of interweaving bands of cells, embedded with reticular and collagen fibres, and inflammatory infiltrates of mononuclear cell and plasma cells. The tumour is highly vascular and contains abnormally dense and irregular blood vessels that leak red blood cells into the surrounding tissue, thereby giving the tumour the characteristic dark colour.

Skin lesions are divided into three stages: patch, plaque, and nodular. The patch stage shows a proliferation of irregular branch blood vessels that may be grouped around normal-looking blood vessels. The patch stage is characterised by the appearance of spindle cells forming bundles in the vascular spaces. There is evidence of extravasation of red blood cells and macrophage, while mitoses and nuclear abnormalities are more profound in both the spindle and endothelial cells. Biomarkers associated with KSHV-associated lesions include tissue-specific markers such as CD34, vascular-endothelial cadherin, endothelial leukocytes adhesion molecule type 1, CD4, CD68, CD14, and PECAM. Others are cytokine-activated ECs.

The classical form of KS is usually found in the lower extremities, while AIDS-associated KS mostly involves other parts of the body. The skin of the face, the extremities, the torso, and the mucus membrane of the oral cavity are mostly affected. Studies reported of gastrointestinal tract involvement among 40 per cent of patients with AIDS-associated KS at the time of initial diagnosis and 80 per cent at autopsy. KS may also involve the lung parenchyma, bronchial tree, and pleural surface.

9.1.2 Primary Effusion Lymphoma

PEL is also called body-cavity-based lymphoma, a rare lymphoma found in HIV-infected patients. It is a unique form of NHL, derived from clonally expanded malignant B cells and present as a lymphomatous effusion tumour containing various body cavities, such as the pericardium, peritoneum, and pleurum. Others have also reported PEL as a solid mass in the lymph nodes and other regions. PEL is aggressive, can progress rapidly, and causes high fatality, with mean survival time for patients with PEL about two to six months. Biomarkers associated with PEL include CD45, activation-associated antigens, clonal immunoglobulin rearrangement, and CD138/syndecan-1. Other biomarkers are the expression of viral proteins Latency-Associated Nuclear Antigen (LANA)-1, vCyclin, vFLIP, kaposin, and LANA-2. KSHV is detected as either monoclonal or oligoclonal episomes in PEL samples. KSHV genome is found in PEL in high copy number, about 50–150 viral genome per infected cells, and can be used as diagnostic criteria for this lymphoma.

9.1.3 Multicentric Castleman's Disease

MCD is a localised lymphoproliferative condition which is characterised by expansion of germinal centres with B-cell and vascular proliferations. There are different types of MCDs, with the plasmablastic variant form more commonly seen in AIDS patients and transplant recipients. The plasmablastic MCD can be aggressive and rapidly progress to high fatality. MCD is frequently but not always associated with KSHV infection. Biomarkers associated with KSHV-associated MCD include dysregulated high levels of IL-6 and VEGF. KSHV is also detectable in almost all HIV-positive MCD cases and about 50 per cent of HIV-negative MCD cases.

KSHV is postulated to be the cause of other diseases due to dysfunction of the immune system; for example, a newly characterised KSHV-associated condition abbreviated as KICS (KSHV inflammatory cytokine syndrome) has been reported in patients with HIV and KSHV co-infection, displaying elevated levels of IL-6 production. Additionally, KSHV has been linked to different forms of lymphomas, including Burkitt's lymphoma, multiple myeloma, germinotrophic lymphoproliferative disorder (GLD), malignant skin tumours, angio-sarcomas, angio-immunoblastic lymphoma, and primary pulmonary hypertension. Others have also reported of a KSHV/HHV8-associated germinotrophic lymphoproliferative disorder in HIV-seronegative individuals. More studies are needed to analyse the causal association between KSHV and other diseases since there is limited evidence to strongly link KSHV to other diseases such as Kikuchi's diseases, saliva gland tumours, and so on.

9.2 KSHV Genome

Analysis of DNA extracts from purified KSHV showed that the full-length genome is 165 to 170 kb. Moore et al. performed the primary characterisation of the genome. The genome of the virus is similar to that of herpesvirus saimiri in that it has a single contiguous region of 140 to 145 kb which contains all the coding regions. Studies have shown permissive and non-permissive tumour cell lines contain KSHV DNA up to 270 kb in size. The genome has repeats of 803 bp in length, of which over 85 per cent are guanidine and cytosine (G+C). The genome is surrounded by icosahedral protein capsid, thick tegument, and a lipid bilayer envelope. A mature KSHV contains at least twenty-four virus-associated proteins: five capsid proteins, eight envelope proteins (glycoprotein), six tegument proteins, and five proteins with unidentified location. The KSHV gene seems to have circular conformation; however, active DNA found in lytic replication is linear. Close to a hundred genes/ORF (figure 1) encoded by 140 kb long unique region (LUR) with 53.5 per cent G+C content have been described. Many of them are conserved in most herpesvirus. The LUR is about 138 to 140.5 kb long and contains all of KSHV ORFs, which are designated K1 to K15, depending on their relative location in the KSHV genome (normally from left to right). In addition, KSHV contains many genes which originated from the host genome and are homologous of cellular genes. A number of the genes play a significant role in the pathogenesis of KS. Some, such as K1 and K15, are involved in signal transduction, cell cycle regulation by vCyclin, inhibition of apoptosis by vFLIP, and immune modulation by viral chemokines receptors.

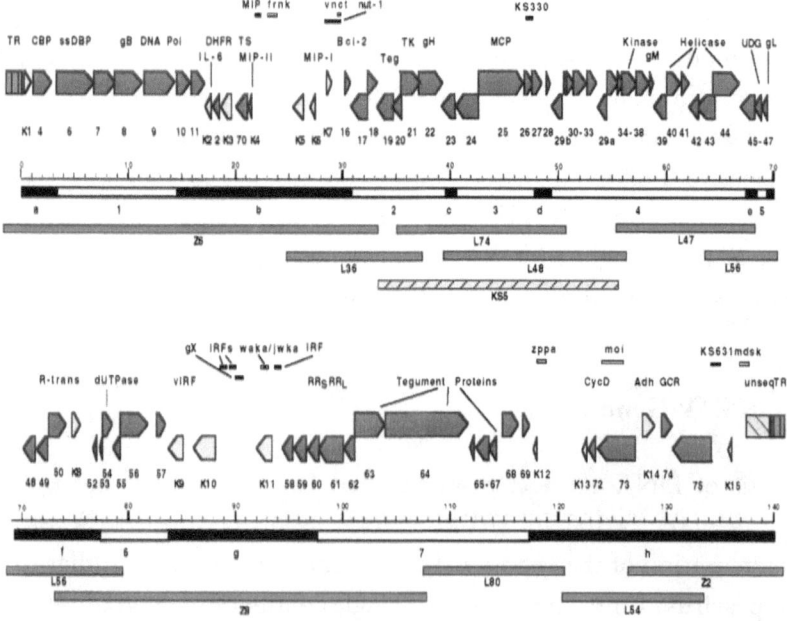

Figure 9.1: KSHV genome description (Adopted from Russo et al., 1996)

In addition, a total of twelve microRNAs have been reported in KSHV genome. Ten were found in the non-coding region between K12/kaposin and K13/orf71/vFLIP, and two were found within K12 ORF. All of them were expressed during the period of latency, with a subset of them upregulated during the lytic cycle. The viral genes encoded by KSHV are classified into three groups: 1. Herpesvirus common genes, 2. KSHV unique genes, and 3. Cellular homologues genes, which may contain group 1 and 2. There are some gaps in our understanding of herpes viruses; further studies are needed to identify genes associated with KSHV pathogenesis.

9.3   KSHV Life Cycle

To initiate its replication (figure 2), the virion particle binds to the host cell surface receptor and penetrates into the host cell cytoplasm via a complex multistep process. KSHV has two different phases in its life cycle. The latent phase is characterised by a circular episome tightly packed as nucleosome and expression of small subset of latent transcripts in the infected cells. The circular episome is chromatinised as a result of its association with cellular histones in order to ensure protection of viral DNA ends to escape the host DNA damage response; stable maintenance, replication, and segregation of the viral genome to daughter cells during mitosis; successful completion of viral life cycle; and regulation of viral gene expression. There is no production of functional viral particles during this phase. OrfK12/Kaposin, orf71/K13/vFLIP, orf72/vCyclin, and orf73/ LANA have been detected in the latent phase of all the KSHV-associated cancers. The second phase is the lytic phase, which is characterised by the replication of linear viral genomes and the expression of more than eighty transcripts in a highly planned temporal order of immediate (IE), early, and late groups.

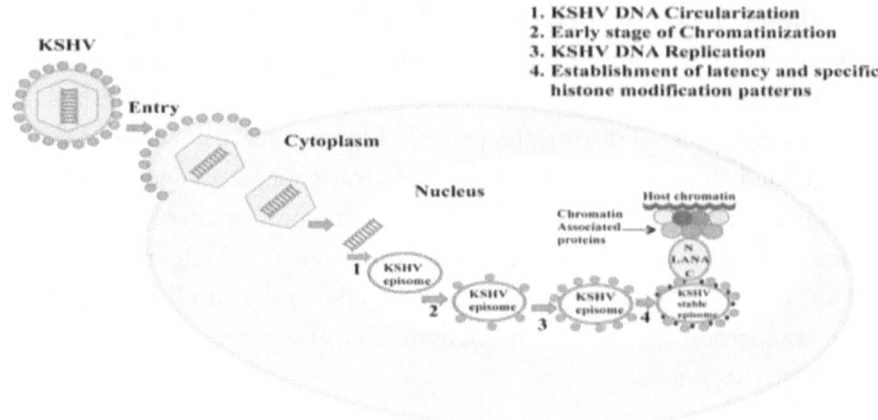

Figure 9.2: KSHV life cycle (Adopted from Uppal et al., 2015)

The IE gene does not rely on viral protein synthesis, but it's important for regulating transcriptional cascade. A KSHV-encoded IE gene, Rta, is required and sufficient for initiating the lytic replication cycle to completion, while another IE gene, orf45, is essential for the suppression of IFN induction during lytic viral infection or reactivation. The general function of the early and late genes is to facilitate the replication of viral genome, viral assemble, and egress.

Although we are limited in our understanding of some aspects of the viral life cycle, studies indicate that there are key regulatory steps in the life cycle of many viruses that are essential for the establishment and persistence of viral latency. After entering the host cell nucleus, the viral genome must adopt a structure that is similar to the host genome and interact with cellular chromatin. The chromatinisation of viral DNA is influenced by the same epigenetic factors as cellular DNA, resulting in the generation of a viral epigenome which has an essential role in both latency and lytic reactivation of the viral genome. In addition, some studies have shown that when viruses enter and hide in the host nucleus, they co-evolve with numerous cellular chromatin modulation mechanisms, which ensures their survival and propagation.

9.4 Viral Genes Involved in KSHV-Associated Transformation and Oncogenesis

A number of genes are involved in lytic and latent cycle replications which are essential for the long-term persistence of the virus. These gene products contribute to KSHV-induced pathogenesis. This section will review some of the genes associated with KSHV oncogenesis.

9.4.1 Latency-Associated Nuclear Antigen

LANA is a 222-232 kDA nuclear protein that tethers the viral episomal DNA to cellular chromosome through histone HI binding. It is the main latent protein of KSHV, which is expressed in all types of KSHV-associated tumours and is the key viral protein associated with viral oncogenesis. LANA is a transcriptional regulator that suppresses KSHV Rta expression, resulting in the inhibition of viral lytic replication and maintaining latency. LANA is highly promiscuous in its transforming activities; it interacts with many other cellular proteins and disrupts a number of cellular proliferation control mechanisms by binding to glycogen synthese kinase 3β (GSK-3β), a signal protein in the Wut pathway, and negatively regulating β-catenin, thereby increasing the levels of β-catenin and activity of downstream transcription factor TCF/LEF. Other reported targets include p53, pRB, AP-1, STAT, and p300.

9.4.2 V-Cyclin

This homolog of cellular cyclin D2 is encoded by ORF72. It is known to activate cellular cyclin-dependent kinase 6 (CDK6) to regulate cellular proliferation and viral replication, promotes G1-S transition of the cell cycle, apoptosis, induce DNA damage, has oncogenic potential and autophagic properties and activates NF-kβ. It is expressed during KSHV's latency and lytic replication phases. The v-cyclin-CDK6 complex can phosphorylate pRB protein. The exact role of this protein in KSHV replication is still not fully understood, but studies indicate that v-cyclin-CDK6 complex mediates the phosphorylation of nucleophosmin (NPM), which facilitates NPM-LANA interaction as well as recruitment of HDAC1, resulting in KSHV latency.

Furthermore, studies have shown that v-cyclin shares a close functional relationship with murine gammaherpesvirus 68 v-cyclin, which is known to mediate efficient lytic reactivation from latency. A study also showed that in vivo expression of v-cyclin in B- and T-cell lymphocytes led to markedly lower survival due to the frequency of early-onset T-cell lymphoma and pancarditis. Finally, a study suggested that cyclin can play a role in the initiation of Notch-dependent lymphomagenesis because the Notch pathway is known to have a role in T-cell development and lymphoma initiation.

9.4.3 v-FLICE (fas-Associated Death Domain-Like il-1 β-Covertase Enzyme) Inhibitory Protein (vFLIP)

vFLIP, also known as K13, is encoded by ORF17. The FLIP proteins are a group of cellular and viral proteins that inhibit death receptor DR-induced apoptosis. They are made of two death effector domains (DEDs) capable of inhibiting DED-DED interaction between FAS-associated protein with death domain (FADD) and procaspases 8-10 with death signaling complex (DISC). VFLIP is latently expressed through splicing of LANA transcript from messenger RNA and through the use of IRES in v-cyclin coding sequence. In KSHV, vFLIP prevents recruitment and processing of procaspases 8, thereby inhibiting FAS-induced apoptosis, thus providing a survival advantage for KSHV-infected cells. Another function of vFLIP is it binds to IkB kinase (IKK) complex and heat shock protein 90 (hsp90), resulting in both classical and alternative NF-kB survival signaling. Elimination of vFLIP by RNA interference results in an appreciable decrease in NF-kB activity and apoptosis. This confirms that vFLIP has a role in the survival of PEL cells. More studies are needed to understand the role of vFLIP in the initiation of KSHV-associated diseases.

9.4.4 Kaposin (Kpsn)

Kpsn is the most abundantly expressed viral transcript during KSHV latency. Three types of Kpsn have been described: Kpsn A, Kpsn B, and Kpsn C. Kpsn A has oncogenic properties and transforms cells in culture. A study reported of morphological change in Rat-3 cells through the interaction with cytohedsin-1. Kpsn B increases the expression of cytokine by blocking the degradation of mRNAs, thereby stabilising cytokine expression such as IL-6 and GM-CSF. mRNA stabilisation activity depends on direct repeats (DR1 and DR2) of elements of Kpsn B. A study has found that Kpsn B is abundant in PEL cell line BCBL-1. The function of Kpsn C is not known.

9.4.5   Replication and Transcription Activator (Rta)

Rta is a novel E3 ubiquitin ligase which is encoded by ORF50. It targets a number of transcriptional repressor ubiquitin proteasome pathways. A study showed that Rta interacts with the cellular transcriptional repressor protein Hey1 and that Hey1 has a contributory role in the maintenance of KSHV latency, although other studies showed that Rta results in the disruption of latency. In addition, Rta is reported to be important in the induction of KSHV lytic replication, from latency through the activation of the lytic cascade.

9.4.6   KSHV MicroRNAs (MiRNAs)

MiRNAs are non-coding RNAs of nineteen to twenty-three nucleotides in length with the ability of regulating gene expression post-transcriptionally by targeting 3' untranslated regions (UTRs) of messenger RNAs. A number of cellular targets of KSHV-encoded miRNAs have been identified. They are highly expressed in latency and KS tumours, which shows that they have essential functions in the viral life cycle and development of KS tumours. Their roles include inhibition of apoptosis and transformation, cell cycle regulation, and promotion of angiogenesis. Earlier, it was suggested that these viral post-transcriptional regulators might be promoters of latency by targeting lytic genes. In a study profiling mRNA mimic and antagomir, Ziegelbauer et al. reported that KSHV miRNAs can modulate the latent/lytic transition through direct targeting of RTA by miR-K12-9. Other studies showed that miR-K12-5 and miR-K12-7 can directly target RTA. This data is consistent with in silico prediction of targets within RTA 3'UTR. Targeting of RTA could prevent reactivation. MiRNAs can also contribute to latency by targeting host factors that are involved in viral reactivation. A study showed that KSHV miRNAs play a role in maintaining latency by targeting cellular transcription factors and that miR-k12-11 and miR-k12-3 prevent lytic reactivation through modulating the expression of transcription factors MYB, C/EPPα, and Ets-1. Moody et al. found in a study that miRNA redundantly targets the NF-kB pathway to regulate cell cycle progression and apoptosis.

9.4.7   K1

K1 is a 46 kDA type 1 membrane glycoprotein encoded by ORFK1, which is the most variable portion of the viral genome. It is also referred to as variable ITAM-containing protein (VIP) because it contains an immunoreceptor tyrosine-based activation motif (ITAM). KSHV K1 ITAM activates several intracellular signaling pathways, especially P13K/AKT. This means the expression of K1 results in the inhibition of proapoptotic proteins and increases the life span of KSHV infected cells. Additionally, K1 enhances the production of inflammatory cytokines and proangiogenic factors such as vascular endothelial growth factor. KSVH K1 immortalises primary human umbilical vein endothelial cells (HUVEC) in culture and transforms rodent fibroblasts as well as inducing tumour in vivo in transgenic mice. K1 has been detected in KS, PEL, and MCD. Available data suggests that K1 is essential in KSHV-associated tumorigenesis and angiogenesis.

9.4.8   Viral Interleukin (vIL-6)

vIL-6 is encoded by ORFK2 and is homolog to cellular IL-6. Long before the discovery of KSHV, vIL-6 was suspected to be important in the pathogenesis of KS and MCD. Data shows that vIL-6 mimics some IL-6 activities, such as stimulating growth of IL-6 dependent cells and triggering JAK, STAT3, MAPL, and H7-sensitivity pathways. The JAK and STAT3 pathways stimulated by vIL-6 result in increased VEGF expression. However, there are some differences in receptor usage which may give rise to the underlying quantitative and qualitative difference in the utilisation of signal pathways. While cellular IL-6 depends on IL-80 and gp 130, vIL-6 signals can be attained through gp130 alone. vIL-6 is capable of inducing transcriptional activation through Type IL-6 response elements (REs) that bind C/EBP. This is indicative of Ras-MAP kinase pathway induction. From available data, it can be suggested that vIL-6 contributes to the progression of KSHV-associated diseases by continued activation of IL-6 stimulated growth and anti-apoptotic pathways; for example, a study found that Castleman's disease involved aberrant IL-6 activity from either endogenous or viral sources.

9.4.9   Viral Interferon Regulatory Factors (vIRFs)

vIRFs are a family of genes with homology to cellular interferon regulatory factors (IRFs) .There are four types of vIRFs, which inhibit the activity of their cellular counterparts. vIRF-1 downregulates interaction with cellular p53 via its central DNA domain. This interaction inhibits transcriptional activation of p53. A study by Shin et al. showed that KSHV vIRF-1 downregulates the total p53 protein level by facilitating its proteasome-mediated degradation. vIRF-1 interacts with cellular ataxia telangiectasis-mutated (ATM) kinase via its carboxyl-terminal transactivation domain. This interaction blocked the activation of ATM kinase activity induced by DNA damage stress. As a result, vIRF-1 expression greatly reduced the level of serine phosphorylation of p53. It leads to increase of p53 ubiquitination and decrease of its protein stability. The study indicated that KSHV vIRF-1 greatly compromised the ATM/p53-mediated DNA damage response checkpoint, as it targets both upstream ATM and downstream p53 tumour suppressors. This helps it evade the host growth surveillance and facilitate viral replication in infected cells. vIRF-1 and 2 also directly interact with cellular IRFs. In addition, vIRFs have other functions such as modulation of Myc, transforming growth factor-β, and NF-kβ signaling. These activities of vIRFs have been implicated in KSHV tumorigenesis.

9.4.10   Viral G Protein-Coupled Receptor (vGPCR)

vGPCR is encoded by ORF74. It is a lytic cycle-associated protein, highly angiogenic, and homologue of IL-8 receptor that signals in part via the cytoplasmic protein tyrosine phosphatase Shp2. It induces several signaling pathways leading to the activation of various transcription factors and ultimately leading to the expression of cellular and viral genes involved in survival, proliferation, and angiogenesis. The expression of vGPCR has been reported in small fractions of KS, PEL, and MCD. The role of this protein in KSHV tumorigenesis has been well documented. A vGPCR transgenic mice developed KS-like angioproliferative lesions with surface markers and cytokine profiles resembling those of KS. It has been shown that in KSHV-associated malignancies, the expression of vGPCR was found in some cells in transgenic tumours and a few other tissues, which suggests that vGPCR-mediated tumour formation was driven by spontaneous lytic reactivation in the background of latently infected cells. VEGR was reported to be increased in vGPCR-induced tumours. vGPCR can transform NIH3T3 fibroblasts in vitro, and vGPCR-expressed 3t3 cells form tumours in mice. vGPCR also activate mitogen-activated protein kinase (MAPKs), p13K, and Akt in endothelial cells, and all these signaling pathways leads to the activation of key cellular transcription factors, such as activating protein-1 (AP-1), NF-kB, nuclear factor activator of T cells, cyclic AMP response element binding protein (CREP), and hypoxia-inducible factor-1 (HIF-1). These transcription factors in turn regulate several viral genes such as vIL-6. Just like K1, autocrine/panacrine signaling of vGPCR might have a role in KSHV-associated oncogenes and angiogenesis.

9.4.11   mRNA Transcript Accumulation (MTA)

MTA is encoded by ORF57 protein. It is made of nuclear protein composed of 455 amino acid (aa) residues. Motif analysis of KSHV ORF57 aa sequence showed several sequence motifs which remotely resemble those found in cellular RNA binding proteins. Two simple RGG motifs have been described: RGG1, composed of 138-140 aa, and RGG2, made of 372-374 aa. The RGG motifs are similar to RGG-box of RNA-binding proteins: serine/arginine or arginine/serine dipeptides made of 77-95 aa, a nonconsensus putative adenine-thymine (AT) hook made of 119-130 aa, a putative leucine-rich region of 343-364 aa, and a γ-herpesvirus glycine/phenylalanine/phenylalanine (GLFF) motif made of 448-451 aa, whose function is unknown.

Furthermore, the N-terminal half of KSHV ORF57 is enriched with polar residues to form short acidic regions made of approximately 7-52 aa, followed with a basic region made of high content of arginine residues that harbour all functionally redundant nuclear localisation signals (NLSs), of which three forms have been described: NLS1, made of 101-107 aa; NLS2, made of 121-130 aa; and NLS3, made of 143-152 aa. KSHV ORF57 is essential for efficient expression of KSHV lytic genes and replicative replication. Deletion of this protein from the virus genome resulted in inefficient expression of viral lytic genes and abortive viral replication.

It possesses a number of activities that are essential for the expression of viral genes, including the three major functions of enhancement of RNA stability, promotion of RNA splicing, and stimulation of protein translation. MTA is multifunctional and can interact with a number of cellular cofactors. These interactions are essential for the formation of MTA-containing ribonucleoprotein complexes, which are specific binding sites in the target transcripts (referred to as MTA-responsive elements [MREs]).

Two structurally distinct domains have been identified within ORF57 polypeptide: a structure α-helix rich c-terminal and an unstructured intrinsic disordered N-terminal domain. These distinctive structures allow for their unique binding affinities, of which N-terminal domain mediates the interaction with cellular cofactors and target RNAs; the C-terminal contributes to the stability of ORF57 protein in infected cells by counteracting caspase- and proteasome mediation degradation pathways. Although our knowledge of KSHV ORF57 has improved over the past years, there are still some gaps in the data on the biochemistry and biophysics properties of individual domains of this protein; therefore, future studies should target these areas.

9.4.12   Viral Processivity Factor (vPF)

vPF is encoded by ORF59 and is one of the factors to be recruited by RTA to oriLyt, where it acts as an accessory factor or sliding clamp, stabilising the binding. In a study, Rossetto et al. showed that binding ORF59 to the C/EBPα binding motif within oriLyt is essential for its function and is dependent on the presence of RTA. This means the function of viral polymerase is also dependent upon its interaction. ORF 59 forms a homodimer, which translocates viral polymerase, which is encoded by ORF9 into the nucleus for efficient synthesis of DNA fragments. This is the processivity function. ORF59 is a phosphoprotein that is phosphorylated by KSHV viral Ser/Thr kinase (ORF36). In a study, McDowell et al. showed that ORF36 phosphorylates ORF59 at Ser 378, which is essential for ORF59's ability to bind RTA and oriLyt. In addition, replacing the phosphorylated serine of these sites within alanines critically reduces viral products. However, the precise mechanism by which ORF59 interacts with RTA has not been elucidated. Identifying this mechanism will help develop a strategy to regulate viral lytic DNA replication at oriLyt.

9.4.13   KSHV bZIP (K-bZIP)

Also referred to as K8, this is an immediate early protein which overlaps with ORF50 (Rta) and requires splicing. It can also activate twenty-one KSHV promoters. K-bZIP gene locus is made of and controls two promoters: one early controlling, K-bZIP, and one late controlling, K8.1. K-ZIP, which can be homodimerised consisting of 237 amino acids. It contains several functional domains: a transcription activation domain at the N terminus, a SUMO interaction motif, a leucine zipper domain at its C terminus, a nuclear localisation signal, a DNA binding domain, and a basic region. K-bZIP can be SUMOylated at leucine, and this process can affect its interaction with many cellular and viral proteins.

A number of proteins are known to interact with K-bZIP, including p53, cAMP-response element-binding protein (CREB)-binding protein (CBP), CCAAT/enhancer-binding protein α, and others. The effect of such interaction on gene regulation is either positive or negative for viral growth. K-bZIP represses the ORF activities of transduction, which suggests that it has a repressing effect on viral gene expression and viral replication. On the other hand, knockdown of K-bZIP either abolishes the reactivation of KSHV, implying that it is essential for KSHV lytic infection, or lowers viral DNA copies at the latent phase of viral infection, suggesting that K-bZIP might have a role in abortive lytic DNA replication of de novo infection or the maintenance of latent viral genome. In a study, Martinez and Tang found that K-bZIP interacts and colocalises with histone deacetylase (HDAC) 1/2 in the DNA replication domain. This means K-bZIP might function through either recruiting HDAC to have a negative effect on some genes regulation or segregating HDAC and inhibiting its activity, thereby having a positive effect on gene regulation. They also discovered that leucine zipper domain is required by K-bZIP to interact with HDAC 1/2 and some other KSHV lytic gene promoters. These interactions are essential for KSHV to replicate HER 293T cells.

9.5   KSHV Oncogenesis

As explained earlier, the life cycle of KSHV consists of latent and lytic replication phases, which are important for the development of KS tumours. Therefore, understanding the mechanisms that result in latency and reactivation holds the key to elucidating KSHV-associated pathogenesis and devising better therapeutic strategies. This section will review recent advances in mechanisms essential in the KSHV life cycle.

9.5.1   Mechanism of KSHV Latency and Pseudo-Latency

A common feature of all human oncoviruses is that they are persistent latent or pseudo-latent infections that do not generally replicate to form infectious particles in tumours. The virus particle is a naked nucleic acid, often a plasmid or episome, that relies on host cell machinery to replicate. About ninety KSHV genes are expressed during KSHV-associated latent infection. LANA-1, vFLIP, and vCyclin are expressed in latent infections and are found adjacent to one another in KSHV genome, belonging to the multicistronic transcriptional unit referred to as latent transcript (LT) cluster. It has been suggested that LANA-1 might be the principal translation product of longer miRNAs, while vFLIP and vCyclin are produced from shorter transcripts. The three genes are separated from K12 genes, which are usually expressed in low levels during latency, by an approximate 4.5 kb KSHV sequence that lacks significant ORFs. This represents the largest coding gap in the unique region of the KSHV genome. Ten of the KSHV miRNAs are found in this coding gap, while miR-K10 is found within K12 and miR-K12 is found within 3'-UTR. KSHV miRNAs are all expressed in latently infected cells and largely uninfected after induction of lytic replication. KSHV K15 is detected in latently infected cells, but the expression levels increase following induction of lytic replication. K15 is found adjacent to the terminal repeats and is transcribed in a leftward orientation from the terminal repeats region. LANA-2 is abundantly expressed in the nuclei of cultured KSHV-infected PEL cells.

9.5.2 Mechanism of KSHV Persistence

For KSHV to persist, the viral genome must replicate and segregate to progeny nuclei with each cell division. LANA-1 is one of the few genes associated with episomal persistence in the absence of other viral genes. It has been reported that both the N-terminal (N-LANA) and C-terminal (C-LANA) are essential for this function, with the C-terminal suggested to play a supportive role in binding KSHV episomes to host chromosomes. The C-terminal is important for LANA oligomerisation, and data shows that oligomerisation is essential for efficient tethering of KSHV episome. The N-LANA interacts with mitotic chromosomes by binding histones H2A/H2B, while C-LANA binds to KSHV terminal repeats DNA simultaneously.

With this, LANA tethers the viral genome to the host chromosome and distributes viral DNA to daughter cells during mitosis. A study by Sun et al. found that LANA recruits replication factor C(RFC), proliferating cell nuclear antigen (PCNA) loader, and a DNA polymerase damp to drive DNA replication efficiently. They therefore suggested that PCNA loading is a rate-limiting step in DNA replication that is incompatible with viral survival and that LANA enhancement of PCNA loading allows for efficient viral replication and persistence.

9.5.3 Mechanism of KSHV Reactivation

Switching from latent to lytic infection by KSHV is mediated by a number of stimuli that induce the expression of RTA. This expression of RTA is essential and sufficient to trigger the process of lytic infection, which results in the orderly expression of viral proteins, release of viral progeny, and host cell death. The expression of RTA precedes the expression of all other cycloheximide-resistant IE genes and cycloheximide-sensitive early genes. It has also been reported that RTA activation results in complete replication of nascent. After primary infection, KSHV established latent infections, in which few genes are transcribed and expression of RTA tightly repressed. However, studies showed that cloned promoter region of RTA has high basal activity, which indicates epigenetic change and chromatin modeling of KSHV genes, which may be involved in this repression process. Epigenetic changes like DNA methylation play a role in regulating gene expression in normal mammalian development and cancer by transcriptionally silencing key growth regulator. A study by Chen et al. suggested that a mechanism of hypermethylation in RTA protein may regulate its expression and subsequent KSHV reactivation from latency.

In addition, other chromatin modifications, such as histone deacetylation, affect the local chromatin structure, which coupled with DNA methylation may regulate RTA gene transcription. It has been suggested that methylation of RTA promoter region during latency promotes the association of transcriptional repressors and HDAC. Lytic replication can be induced by chemicals such as butyrate, an inhibitor of HDAC, and 12-0-tetradecanoylphorbol-13-acetate (TPA), an inducer of histone acetylases (HAT). Both inducers have the ability to affect the acetylation state of RTA promoter that is dependent on methylation. It can be argued from such data that the control of latency of lytic switching is a function of the chromosome architecture, which will involve interplay between viral RTA and the host factors that regulate chromatin methylation and acetylation. There has been the suggestion that apart from RTA regulation in KSHV reaction, other cofactors are involved; for example, a study suggested that hypoxia is a possible cofactor for KSHV reaction based on clinical observation, which showed that KS tumours often appear on body parts such as feet and arms, where blood and oxygen supply are low compared to other parts. In support of this suggestion, a study reported that hypoxia could induce KSHV lytic replication in PEL cells. Hypoxia induces the accumulation of hypoxia-inducible factors (HIF1/2). In the

KSHV genome, promoters of RTA and ORF34 are reported to contain functional hypoxia response elements (HREs). In addition, a number of studies suggested that specific cell cycle phase and cell differentiation could regulate KSHV reactivation.

9.6   Treatment of KSHV-Associated Malignancies

A wide range of treatment for KS is available, but the treatment options are based on the severity of the disease, KS subtype, and the immune status of the affected. All patients with AIDS-associated KS should receive highly active antiretroviral therapy (HAART). Studies have shown that an effective treatment regimen using antiretroviral agents results in lower incidence of AIDS-associated KS, regression in size and numbers of existing lesions, as well as histological regression of existing KS lesions. Several antiviral agents such as ganciclovir, foscarnet, and cidofovir inhibit KSHV replication in vitro, but some data showed that antiviral therapy with, for example, cidofovir aimed at KSHV did not have any effect by itself for treatment of KS. This may be attributed to a small amount of lytic KSHV in KS tumour. Antiviral agents may therefore be effective as adjunct to more conventional chemotherapy, with liposomal anthracyclines such as doxorubicin and daunorubicin as first-line regimen and paxlitaxel as second-line regimen. Other chemotherapeutic agents include vinorelbine, IFN-α, and IL-12. In addition, the tyrosine inhibitor imatinib and IL-12 demonstrated some activities against AIDS-associated KS.

For PEL, there is no clear established standard of care, and due to its low incidence, randomised clinical studies are presently not feasible. PEL patients have poor prognosis with a median survival of only two to three months after diagnosis. As in KS, patients co-infected with AIDS would benefit from HAART, as spontaneous regression has been reported. Conventional CHOP-like regimen (cyclophosphomide, doxorubicin, vincristine, and prednisolone) did not improve median survival in comparison to other HIV-associated NHL. For HIV-negative cases of PEL, liposomal anthracycline (with or without bortezomib and prednisolone) can be given. Bortezomib, a proteosome inhibitor used for multiple myeloma, has been shown to enhance the in vitro cytotoxic effects of doxorubicin and paclitaxel, and it has been used successfully in combination with anthracycline-based cytotoxic chemotherapeutic combinations. Rapamycin has also shown some promise in treating PEL cells in culture or xenograft model. Radiation therapy can be initiated for patients who cannot tolerate the above treatments options.

Treatment of MCD with HIV infection, HAART is essential, but care should be taken, as life-threatening flares of MCD have been reported due to manifestations of immune reconstitution. Systemic therapy is the backbone of treating patients with MCD, ranging from cytoreduction chemotherapy such as CHOP and ABV (doxorubicine, bleomycin, and vincristine), single agent maintenance chemotherapy, immunomodulating agents, and monoclonal antibodies against IL-6 and CD20 surface markers and inhibitors of KSHV viral replications.

In conclusion, the current treatment strategies employed for KS, PEL, and MCD are suboptimal and have devastating side effects. Targeting signal pathways of these tumours will be ideal strategies resulting in more beneficial treatment options than the conventional chemotherapy. Therefore, more case and randomised clinical studies are needed to advance treatment options for KSHV-associated diseases.

REFERENCES

Ablashi DV, Clatlynne LG, et al (2002), Spectrum of Kaposi's sarcoma-associated herpesvirus, or human herpesvirus 8 disease. Clin Microbiol Rev 15:439–464.

Al Mehairi S, Lerasoli E, Sinclair AJ (2005), Investigation of the multimerisation region of the Kaposi's sarcoma-associated herpesvirus (human herpesvirus 8) protein K-bZIP: The proposed leucine zipper region encodes a multimerisation domain with an unusual structure. J Virol 79:7905–7910.

Amin HM, Medeiro LJ, et al (2002), Dissolution of the lymphoid follicle is a feature of the HHV 8+ variation of plasma cell Castleman's disease. Am J Pathol 27:91–100.

An J, Sun Y, et al (2004), Antitumor effects of bortezomib (PS-341) on primary effusion lymphoma. Leukemia 18: 1699–1704.

Arvanitakis L, Mesri EA, et al (1996), Establishment and characterization of a primary effusion lymphoma (body cavity-based) lymphoma cell line (BC-3) harboring Kaposi's sarcoma-associated herpesvirus (KSHV/HHV-8) in the absence of Epstein-Barr virus. Blood 88:2648–2654.

Ballestas ME, Chatis PA, Kaye KM (1999), Efficient persistence of extrachromosomal KSHV DNA mediated by latency-associated nuclear antigen. Science 284: 641–644.

Ballestas ME, Kaye KM (2001), Kaposi's sarcoma-associated herpesvirus latency-associated nuclear antigen 1 mediates episome persistence through cis-acting terminal repeat (TR) sequence and specifically binds TR DNA. J Virol 75:3250–3258

Ballon G, Chen K, et al (2011), Kaposi sarcoma (KSHV) vFLIP oncoprotein induces B cell transdifferentiation and tumorigenesis in mice. J Clin Invest 121:1141–1153.

Barbera AJ, et al. (2006), The nucleosomal surface as a docking station for Kaposi's sarcoma herpesvirus LANA. Science 311:856–861.

Barrison IG, Foster S, et al (1988), Upper gastrointestinal Kaposi's sarcoma in patients positive for HIV antibody with cutaneous disease. Br Med J 296:92–99.

Bechtel JT, Liang Y, et al (2003), Host range of Kaposi's sarcoma-associated herpesvirus cultured cells. J Virol 77:6474–6481.

Botterro V, Sharma-Walia N, et al (2009), Kaposi sarcoma-associated herpes virus (KSHV) G protein-coupled receptor (vGPCR) activates the ORF 50 lytic switch promoter: A potential positive feedback loop for sustained ORF50 gene expression. Virology 392:34–51.

Cai X, Cullen BR (2006), Transcriptional origin of Kaposi's sarcoma-associated herpesvirus microRNAs. J Virol 80:2234–2242.

Cai X, Lu S, et al (2005), Kaposi's sarcoma-associated herpesvirus expresses an array of viral microRNAs in latently infected cells. PNAS USA 102: 5570–5575.

Carbone A, Cilia AM, et al (1997), Establishment of HHV-8-positive and HHV-8-negative lymphoma cell lines from primary lymphomatous effusion. Int J Cancer 73: 562–569.

Carbone A, Gloghini A, et al (2005), Kaposi's sarcoma-associated herpesvirus/human herpesvirus type 8 in positive solid lymphoma: A tissue-based variant of primary effusion lymphoma, J Mol Diagn 7:12-27. ars: translating pathophysiology to patient care. Brit J Haematol 129: 3–17.

Cesarman E, Chang Y, et al (1995), Kaposi's sarcoma-associated herpesvirus-like DNA sequences in AIDS-related body cavity-based lymphoma. NEJM 32:1186–1191.

Chan SR, Chandra B (2000), Characterization of human herpesvirus 8 ORF59 protein (PF-8) and mapping of the Processivity and viral DNA polymerase-interacting domain. J Virol 74:10920–10929.

Chaudhary PM, Jasmin A, et al (1999), Modulation of the NF-kappa B pathway by virally encoded death effector domain-containing proteins. Oncogene 18:5738–5746.

Chen J, Ueda K, et al (2001), Activation of latent Kaposi's sarcoma-associated herpesvirus by demethylation of the promoter of the lytic transactivation. PNAS USA 98:4119–4129.

Chen Y, Cinstea M, Ricciardi RP (2005), Processivity factor of KSHV contains a nuclear localization signal and binding domain for transporting viral DNA polymerase into the nucleus. Virology 340:183–191.

Chenna TW (2004), AIDS-related cancer in the era of highly active antiretroviral therapy (HAART): A model of the interplay of the immune system, virus, and cancer. Cancer Invest 22:774–786.

Cool CD, Rai PR, et al (2003), Expression of human herpesvirus 8 in primary pulmonary hypertension. NEJM 349: 1113–1123.

Conrad NK, Steitz JA (2005), A Kaposi's sarcoma virus RNA element that increases the nuclear abundance of intronless transcripts. Embo J 24:1831–1841.

Cotter MA, 2nd, Robertson ES (1999), The latency-associated nuclear antigen tethers the Kaposi's sarcoma-associated herpesvirus genome to host chromosomes in body cavity-based lymphoma cells. Virology 264:254–264.

Danzig JB, Brandt LJ, et al (1991), Gastrointestinal malignancies in patients with AIDS. Am J Gastroenterol 86: 715–718.

Davis DA, Rinderknecht AJ, et al (2001), Hypoxia induces lytic replication of Kaposi's sarcoma-associated herpesvirus. Blood 97:3244–3250.

Deloose ST, Smit LA, et al (2003), High incidence of Kaposi sarcoma-associated herpesvirus infection in HIV-related solid immunoblastic/plasmablastic diffuse large B-cell lymphoma. Leukemia 19:851–853.

Deng H, Young H, Sun R (2000), Auto-activation of the rta gene of human herpesvirus-8/ Kaposi's sarcoma-associated herpesvirus: potential role in infection and malignant transformation. J Virol 78:11108–11120.

Direkze S, Laman H (2004), Regulation of growth signaling and cell cycle by Kaposi's sarcoma-associated herpesvirus genes. Int J Exp Pathol 85:305–319.

Dittmer DP, Krown SE (2007), Targeted therapy for Kaposi's sarcoma and Kaposi's sarcoma-associated herpesvirus. Curr Opin Oncol 19:452–457.

Dourmishev LA, Dourmishev AL, et al (2003), Molecular genetics of Kaposi's sarcoma-associated herpesvirus (human herpesvirus-8) epidemiology and pathogenesis, Microbial Mol Biol 67:175–213.

Dublin N, Fisher C, et al (1999), Distribution of human herpesvirus-8 latently infected cells in Kaposi's sarcoma, multicentric Castleman's disease, and primary effusion lymphoma. PNAS USA 96: 4546–4551.

Efklidu S, Bailey R, et al (2008), vFLIP from KSHV inhibits anoikis of primary endothelial cells. J Cell Sci 121:450–457.

Friborg J, Jr, Kong W, et al (1999), p53 inhibition by the LANA protein of KSHV protects against cell death. Nature 402:889–894.

Fujimuro M, Hayward SD (2008), The latency-associated nuclear antigen of Kaposi's sarcoma-associated herpesvirus manipulates the activity of glycogen kinase-3 beta. J Virol 77:8019–8030.

Gaidano G, Cechova K, et al (1996), Establishment of AIDS-related lymphoma cell lines from lymphomatous effusions. Leukemia 10:1237–1240.

Gould F, Harrison SM, et al (2009), Kaposi's sarcoma-associated herpes virus RTA promotes degradation of the Hey repressor protein through the ubiquitin proteasome pathways. J virol 83:6771–6738.

Gruffat H, Portas-Seritis S, et al (1999), Kapos's sarcoma-associated herpesvirus (human herpesvirus 8). Virol 80:557–561.

Guasparri I, Keller SA, Cesarman E (2004), KSHV vFLIP is essential for the survival of infected lymphoma cells. J Exp Med 199:993–1003.

Guito J, Lukac DM (2012), KSHV Rta promoter specification and viral reactivation. Front Microbiol 3:30.

Haque M, Wang V, et al (2006), Genetic organization and hypoxic activation of the Kaposi's sarcoma-associated herpesvirus ORF34-37 gene cluster. J Virol 80:7037–7051.

Han Z, Swaminathan S (2006), Kaposi's sarcoma-associated herpesvirus lytic gene ORF57 is essential for infectious virus production. J Virol 80:5251–5260.

Hocqueloux L, Agbalika F, et al (2001), Long-term remission of an AIDS-related primary effusion lymphoma with antiretroviral therapy. AIDS 15:280–282.

Huang YQ, Friedman-Kien AE, et al (1993), Cultured Kaposi's sarcoma cell lines express factor XIIIa, CD14, and VCAM-1, but not factor VIII or ELAM-1. Arch Dermatol 129:1291–1296.

Lim C, Choi C, Choe J (2004), Mitotic chromosome-binding activity of latency-associated nuclear antigen 1 is required for DNA replication from terminal repeat sequence of Kaposi's sarcoma-associated herpesvirus. J Virol 78:7248–7256.

Irmler M, et al (1997), Inhibition of death receptor signals by cellular FLIP. Nature 388:190–195.

Izumiya Y, Ellison TJ, et al (2005), Kaposi's sarcoma-associated herpesvirus k-bZIP represses gene transcription via SUMO modification. J Virol 79: 9912–9925.

Jacobs SR, Damenia B (2011), The viral interferon regulatory factors of KSHV: Immunosuppressors or oncogenes. Front Immunol 2:19.

Jeong JH, et al. (2004) Regulation and autoregulation of the promoter for the Latency-associated nuclear antigen of Kaposi's sarcoma-associated herpesvirus. J Biol Chem 279:16822–16831.

Johnson AS, Maroman N, Vieira J (2005), Activation of Kaposi's sarcoma-associated herpesvirus lytic gene expression during epithelial differentiation. J Virol 79:13769–13777.

Judith JG, Lacoste V, et al (2000), Monoclonality or oligoclonality of human herpesvirus 8 terminal repeat sequence in Kaposi's sarcoma and other diseases. J Natl Cancer Inst 92:729–736.

Katano H, Sato Y, et al (2000), Expression and localization of human herpesvirus 8-encoded protein in primary effusion lymphoma, Kaposi's sarcoma, and multicentric Castleman's disease. Virology 269: 335–344.

Keller SA, et al (2006), NF-kappa B is essential for the progression of KSHV-EBV-infected lymphomas in vivo. Blood 101:3295–3302.

Keller SA, Schttner EJ, Cesarman E(2007), Inhibition of NF-kappa B induces apoptosis of KSHV-infected primary effusion lymphoma cells. Blood 96:2537–2542.

Piolot T, Tramier M, et al (2001) :Close but distinct regions of human herpesvirus 8 latency-associated nuclear antigen 1 are responsible for nuclear targeting and binding to human mitotic chromosomes. J Virol 75: 3948–3959.

Komenduri KV, Luce JA, et al (1996), The natural history and molecular heterogeneity of HIV-associated primary malignant lymphomatous effusions. J Acquir Immune Defic Syndr Hum Retrovirol 13:215–226.

Kon HB, Bubley GJ, et al (2005), Imatinib-induced regression of AIDS-related Kaposi's sarcoma. J Clin Oncol 23:982–989.

Lacosta V, Nicot C, et al (2007), In primary effusion lymphoma, MYB transcriptional repression is associated with v-FLIP expression during latent KSHV infection while both v-FLIP and vGPCR becomes involved during lytic cycle. Br J Haematol 138:487–501.

Laine L, Amerian J, et al (1990), The response of symptomatic gastrointestinal Kaposi's sarcoma to chemotherapy: A prospective evaluation using an endoscopic method of disease quantification. Am J Gastroenterol 85:959–961.

Liao W, Tang Y, et al (2003), The K-bZIP of Kaposi's sarcoma-associated herpesvirus/human herpesvirus 8 (KSHV/HHV-8) Rta and represses Rta-mediated translation. J Virol 77:3809–3815.

Lim C, Choi C, Choe J (2004), Mitotic chromosome-binding activity of latency-associated nuclear antigen 1 is required for DNA replication from terminal repeat sequence of Kaposi's sarcoma-associated herpesvirus. J Virol 78:7248–7256.

Laurent C, Meggetto F, et al (2008), Human herpesvirus 8 infection in patients with immunodeficiencies. Hum Pathol 39:983–993.

Leads for the MMWR (1982), Classification system for human T-lymphotropic virus type III/lymphadenopathy-associated virus infection. JAMA 256: 20–21, 24–25.

Lee JS, et al (2009), FLIP-mediated autophagy regulation in cell death control. Nat Cell Biol 11: 1355–1362.

Liang X, Padeu CR, et al (2011), Murine gammaherpesvirus immortalization of fetal liver-derived B cells requires both the viral cyclin D homolog and latency-associated nuclear antigen. PLoS Pathog 7:e1002220.

Lieberman PM (2013), Keeping it quiet: Chromatin control of gammaherpesvirus latency. Nat Rev Microbiol 11:863–875.

Lieberman PM (2006), Chromatin regulation of virus infection. Trends Microbiol 14:132–140.

Lin R, Genin P, et al (2001), HHV-8 encoded vIRF-1 represses the interferon antiviral response by blocking IRF-3 recruitment of the CBP/p300 coactivators. Oncogene 20:800–811.

Lin SF, Robinson DR, et al (1999), Kaposi's sarcoma-associated herpesvirus encodes a b2IP protein with homology to BZLFI of Epstein-Barr virus. J Virol 75:1909–1917.

Lin X, Liang D, et al (2011), miR-k12-7-5p encoded by Kaposi's sarcoma-associated herpesvirus stabilizes the latent state by targeting viral ORF 50/RTA. PLos One 6:e16224.

Low W, Harries M, et al (2001), Internal ribosomes entry sites regulated translation of Kaposi's sarcoma-associated herpesvirus FLICE inhibitory proteins. J Virol 75:2938–2945.

Lu F, Stedman W, et al (2010), Epigenetic regulation of Kaposi's sarcoma-associated herpesvirus latency by virus-encoded microRNAs that target RTA and the cellular Rb12-DNMT pathway. J Virol 84:2697–2706.

Lukac DM, Kirshner JR, Ganem D (1999), Transcriptional activation by the product of open reading frame 50 of Kaposi's sarcoma-associated herpesvirus is required for lytic viral reactivation in B cells. J Virol 73:9348–9361.

Majerciak V, Pripuzova N, et al (2007), Targeted disruption of Kaposi's sarcoma-associated herspevirus ORF57 in the viral genome is detrimental for the expression of ORF57, K8 alpha, and K8.1 and the production of infectious virus. J Virol 81:1062–1071.

Majerciak V, Ni T, et al (2013), A viral genome landscape of RNA-polyadenylation from KSHV latent to lytic infection. PLoS Pathog 9: e1003749.

Martinez FP, Tang Q (2012), Leucine sipper domain is required for Kaposi sarcoma-associated herspesvirus (KSHV) K-bZIP protein to interact with histone deacetylase and is important for KSHV replication. J Biol Chem 287:15627–15634.

Nakamura H, Lu M, et al (2003), Global changes in Kaposi's sarcoma-associated virus gene expression patterns following expression of a tetracycline-inducible Rta transactivation. J virol 77: 4205–4220.

Majerciak V, Yamanegi K, et al (2006), Structural and functional analysis of Kaposi sarcoma-associated herpesvirus ORF57 nuclear localization signals in living cells. J Biol Chem 281: 2836–2837.

Majerciak V, Zheng Z-M (2015), KSHV ORF57, a protein of many faces. Viruses 7:604–633.

Matta H, Chaudhary PM (2014), Activation of alternative NF-kappa B pathway by human herpes virus 8-encoded Fas-associated death domain-like IL-1 beta-converting enzyme inhibitory protein (vFLIP). PNAS USA 101:9399–9404.

McDowell M, Purushothaman P, et al (2013), Phosphorylation of Kaposi's sarcoma-associated herpesvirus Processivity factor ORF59 by a viral kinase modulates its ability to associated with RTA and oriLyt. J Virol 87:8038–8052.

Moody R, Zhu H, et al (2013), MicroRNAs mediate cellular transformation and tumorigenesis by redundantly targeting cell growth and survival pathways. PLoS Pathog 9: e1003857.

Montaner A, Sodhi A, et al (2001), The Kaposi's sarcoma-associated herpesvirus G protein-coupled receptor promotes endothelial cell survival through the activation of Akt/protein kinase B. Cancer Res 61:2614–2648.

Murandher S, Pumfery AM, et al (1998), Identification of Kaposin (open reading frame K12) as a human herpesvirus 8 (Kaposi's sarcoma-associated herpesvirus) transforming gene. J Virol 72:4980–4985.

Murphy E, Vanicek J, et al (2008), Suppression of immediate-early viral gene expression by herpesvirus-coded microRNAs: Implication for latency. PNAS USA 105:5453–5455.

Nador RG, Cesarman E, et al (1996), Primary effusion lymphoma: A distinction clinicopathologic entity associated with Kaposi's sarcoma-associated herpesvirus. Blood 88:645–656.

Nakamura H, Lu M, et al (2003), Global changes in Kaposi's sarcoma-associated virus gene expression patterns following expression of a tetracycline-inducible Rta transactivation. J Virol 77:4205–4220.

Oksenhendler E, Clauvel JP, et al (1998), Complete remission of a primary effusion lymphoma with antiviral therapy. Am J Hematol 51:266.

Osborne J, Moor PS, Cheng Y (1999), KSHV-encoded viral IL-6 activates multiple human IL-6 signaling pathways. Hum Immunol 60:921–927.

Palmeri D, Spadavecchia S, et al (2007), Promoter- and cell-specific transcription transactivation by the Kaposi's sarcoma-associated herpesvirus ORF57/Mta protein. J of Virol 81: 13299–13314.

Park J, Seo T, et al (2000), The K-bZIP protein from Kaposi's sarcoma-associated herpesvirus interacts with p53 and represses its transcriptional activity. J virol 24:11977–11982.

Parravicini C, Chandran B, et al (2000), Differential viral protein expression in Kaposi's sarcoma-associated herpesvirus-infected diseases: Kaposi's sarcoma, primary effusion lymphoma, and multicentric Castleman's disease. Am J Pathol 156: 743–749.

Parravicini C, Corbellino PM, et al (1997), Expression of a virus-derived cytokine, KSHV vIL-6 in HIV seronegative Castleman's disease. Am J Pathol 151:1517–1527.

Pearce M, Matsumura S, et al (2005), Transcripts encoding K12, v-FLIP, v-cyclin, and the microRNAs cluster of Kaposi's sarcoma-associated herpesvirus originate from a common promoter. J Virol 79:14457–14464.

Pekkonen P, Jarviluoma A, et al (2014), KSHV viral cyclin interferes with T-cell development and induces lymphoma through cdk6 and Notch activation in vivo. Cell Cycle 13:3670–3684.

Philpott N, Bakken T, et al (2010), The Kaposi's sarcoma-associated herpes virus G protein-coupled receptor contain an immunoreceptor tyrosine-based inhibitory motif that activates Shp2. J Virol 88: 1140–1144.

Plaisance-Bonstaff K, Choi HS, et al (2014), KSHV miRNAs decrease expression of lytic genes in latently infected PEL and endothelial cells by targeting host transcription factors. Viruses 6:4005–4023.

Radkov SA, Kellam P, Boshoff C (2000), The latent nuclear antigen of Kaposi sarcoma-associated herpesvirus targets the retinoblastoma-E2f pathway and with the oncogene Hras transforms primary rat cells. Nat Med 6:1121–1127.

Renne R, Lagunoff M, et al (1996), The size and conformation of Kaposi's sarcoma-associated herpesvirus (human herpesvirus 8) DNA in infected cells and virus. J Virol 70: 8151–8154.

Rivas C, Thlick AE, et al (2001), Kaposi's sarcoma-associated herpesvirus LANA-2 is a B-cell specific latent viral proteins that inhibit p53. J Virol 75: 429–438.

Rossetto CC, Susilarini NK, Pai GS (2011), Interaction of Kaposi's sarcoma-associated herpesvirus ORF59 with oriLyt is dependent on binding with K-Rta. J Virol 85:3833–3841.

Rossetto C, Gao Y, et al (2007), Transcriptional repression of k-Rta by Kaposi's sarcoma-associated herpesvirus k-bZIP is not required for oriLyt-dependent DNA replication. Virology 369:340–350.

Russo JJ, Bohenzky RA, et al (1996), Nucleotide sequence of Kaposi sarcoma-associated herpesvirus (HHV8). PNAS USA 93:14862–14867.

Samols MA, SkalskyRL, et al (2007), Identification of cellular genes targeted by KSHV-encoded microRNAs. PLOS Pathog 3:e65.

Sarek G, Jarviluoma A, et al (2010), Nucleophosmin phosphorylation by v-cyclin-CDK6 controls KSHV latency. PLoS Pathog 6: e1000818.

Seaman WT, Quinlivan EB (2003), Lytic switch protein (ORF50) response element in the Kaposi's sarcoma-associated herpesvirus K8 promoter is located within but does not require a palindromic structure. Virology 320: 72–84.

Seaman WT, Ye D, et al (1999), Gene expression for the ORF50/K8 region of Kaposi's sarcoma-associated herpesvirus. Virology 263:436–449.

Seo T, Park J, et al (2001), Viral regulatory factor 1 of Kaposi's sarcoma-associated herpesvirus binds to p53 and represses p53-dependent transcription and apoptosis. J Virol 75:6193–6198.

Shin YC, Nakamura H, et al (2006), Inhibition of the ATM/p53 signal transduction pathway by Kaposi's sarcoma-associated herpes virus interferon regulatory factor 1. J Virol 80:2151–2266.

Sinclair AJ (2003), bZIP protein of human gamma-herpesvirus. J Gen Virol 84:1941–1949.

Soulier J, Groflet L,et al (1995), Kaposi's sarcoma-associated herpesvirus-like DNA sequences in multicentric Castleman's disease. Blood 86: 1270–1280.

Sousa-Squiavinato AC, Silvestre RN, Elqui De Oliveriro D (2015), Biology and oncogenicity of the Kaposi sarcoma herpesvirus K1 protein. Rev Med virol Doi:10.1002/rmv.1843 [Epub ahead of print].

Stallone G, Schena A, et al (2005), Sirolinus for Kaposi's sarcoma in renal-transplant recipients. NEJM 352:1317–1323.

Sullivan RJ, Pantanowitz L, et al (2008), Epidemiology, pathophysiology, and treatment of Kaposi's sarcoma disease: Kaposi's sarcoma, primary effusion lymphoma and multicentric Castleman disease. Clin Infect Dis 47: 1209–1215.

Sun Q, Tsurimoto T, et al (2014), Kaposi's sarcoma-associated herpesvirus LANA recruits the DNA polymerase clamp loader to mediate efficient replication and virus persistence. PNAS 111: 11816–11821.

Sun R, Lin SF, et al (1996), Polyadenylylated nuclear RNA encoded by Kaposi sarcoma-associated herpesvirus. PNAS USA 93:11883–11888.

Sun SH, Roy D, et al (2007), Rapamycin is efficacious against primary effusion lymphoma (PEL) cell lines in vitro by inhibiting autrocrine signaling. Blood 109:2165–2173.

Swanton C, Mann DJ, et al (1997), Herpesviral cyclin/cdK6 complexes evade inhibition by CDK inhibitor proteins. Nature 390: 184–184.

Szekely L, et al. (1999), Human herpesvirus-8-encoded LNA-1 accumulates in heterochromatin-associated nuclear bodies. J Gen Virol 80:2889–2900.

Tang J, Zhang ZM (2002), Kaposi's sarcoma-associated herpesvirus K8 exons harbours a K8.1 transcription start site. J Biol Chem 277: 14547–14556.

Thomas M, et al (1999), Viral FLICE-inhibitory proteins (FLIPs) prevent apoptosis by death receptors. Nature 388:517–521.

Tomlinson CC, Damania B (2008), Critical role for endocytosis in the regulation of signaling by the Kaposi's sarcoma-associated herpesvirus K1 protein. J Virol 82:6514–6523.

Uppal T, Banerjee S, et al (2014), KSHV LANA: The master regulator of KSHV latency. Viruses 6:4961–4998.

Uppal T, Jha HG, et al (2015), Chromatinization of the KSHV genome during the KSHV life cycle. Cancers 7:112–142.

Wang SE, Wu FY, et al (2003), CCAAT/enhancer-binding protein-alpha is induced during the early stages of Kaposi's sarcoma-associated herpesvirus (KSHV) lytic cycle reactivation and together with the KSHV replication and transcription activator (RTA) cooperatively stimulated the viral RTA, MTA, and PAN promoters. J Virol 77:9590–9612.

Warden C, Tang Q, Zhu H (2011), Herpesvirus BACs: past, present, and future. J Biomed Biotechnol 2011:124595.

Wen KW, Damania B (2009), Kaposi's sarcoma-associated Herpesvirus (KSHV): Molecular biology and oncogenesis. Cancer Lett doi:10.1016/j.canlet2009.07.004.

Weninger W, Partanen TA, et al (1999), Expression of vascular endothelial growth factor receptor-3 and podoplanin suggest a lymphatic endothelial cell origin of Kaposi's sarcoma tumor cells. Lab Invest 79:243–251.

White DA (1990), Pulmonary complications of HIV-associated malignancies. Clin Chest Med 17: 755–761.

Wong LY, Matchett GA, Wilson AC (2004), Transcriptional activation by the Kaposi's sarcoma-associated herpesvirus latency-associated nuclear antigen is facilitated by an N-terminal chromatin-binding motif. J Virol 78:10074–10085.

Yang TY, Chen SC, et al (2000), Transgenic expression of the chemokines receptor encoded by human herpesvirus 8 induces an angioproliferative disease resembling Kaposi's sarcoma. J Exp Med 191:445–454.

Ye F, Lei X, Gao S-J (2010), Mechanism of Kaposi's sarcoma-associated latency and reactivation. Advances in Virol doi:10.1155/2011/193860.

Ye FC, Zhou FC, et al (2008), Kaposi's sarcoma-associated herpesvirus latent gene vFLIP inhibits viral lytic replication through NF-kappaB-mediated suppression of AP-1 pathway: A novel mechanism of virus control of latency. J Virol 82:4235–4249.

Yu F, Feng J, et al (2007), B cell terminal differentiation factor XBP-1 induces reactivation of Kaposi's sarcoma-associated herpesvirus. The FEB Letters 581:3485–3488.

Yu F, Harade JN, et al (2007), Systematic identification of cellular targets signals reactivating Kaposi sarcoma-associated herpesvirus. PLoS Pathog 3:e44.

Verma D, Li DJ, et al (2015), Identification of the physiological gene targets of the essential lytic replicative Kaposi's sarcoma-associated herpesvirus ORF57 protein. J Virol 89:1688–1702.

Verma SC, Lan K, Robertson E (2007), Structure and function of latency associated nuclear antigen. Curr Top Microbiol Immunol 312:101–136.

Zhi H, Zahoor MA, et al (2015), KSHV vcyclin counters the senescence/G1 arrest response triggered by Nk-kB hyper-activation. Oncogene 34:496–505.

Zhu FX, King SM, et al (2002), A Kaposi's sarcoma-associated herpesviral protein inhibits virus-mediated induction of type 1 interferon by blocking IRF-7 phosphorylation and nuclear accumulation. PNAS USA 99:5573–5578.

Ziegelbauer JM, Sullivan CS, et al (2009), Tandem array-based expression screens identify host mRNA targets of virus encoded microRNAs. Nat Genet 41:130-134.

CHAPTER 10

APPLICATION OF FLOW CYTOMETRY IN ANTI-ONCOVIRAL DRUG DEVELOPMENT

Introduction

Flow cytometry (figure 1) rapidly measures the specific characteristics of a large number of individual cells. Before flow cytometric analysis, cells in suspension are fluorescently labeled, mostly with fluorescently conjugated monoclonal antibody (mAb). In a flow cytometer, the suspended cells pass through a flow chamber (at a rate of 1,000 to 10,000 cells per minute) through the focused beam of laser. After fluorescent activation of the fluorophore at an excitation wave length, a detector processes the emitted fluorescence and light-scattering properties of each cell. Scientists at Naval Research Laboratory developed a flow cytometry technique to detect the presence of viruses in cells and study their growth. It targets the viral RNA. The technique, referred to as locked nucleic acid (LNA) flow cytometry-fluorescence in situ hybridisation (flow-FISH), involves binding of an LNA probe to viral RNA. The researchers demonstrated that the combination of LNA probes with flow FISH can be used to quantify viral RNA in infected cells. This will also allow researchers to monitor changes in viral RNA accompanying antiviral drug treatment.

A number of improved flow FISH methods have been developed. These include partial automation procedure and multicolour flow FISH, which is the current fastest and most sensitive method available. A step-by-step procedure of flow FISH as described by Baerlocher et al., based on their study to measure the average length of telomere, is as follows: Each flow FISH experiment begins with acquisition of premixed calibration (MESF) beads. Several thousand events are collected, and the main fluorescence and coefficient of variation (CV) of each of the five peaks is recorded and plotted against MESF content provided by the manufacturer, to control for the linearity of the instrument. The next steps are related to selecting the optimal values for the detectors, amplifiers, fluorescence compensation setting, and threshold values for analysis of flow FISH samples.

Once an appropriate instrument setting has been selected, it can be saved and recalled for future study, although minor day-to-day adjustments are needed between experiments and between samples. The instrument settings are further adjusted to provide a good separation of the events of interest over the entire range in the selected channels. Various compensation settings are selected for the analysis of cells simultaneously labeled with fluorescein, phycoerythrin (PE), LDS751, and cy-5. Except for the compensation setting of fluorescence 2 channel (FI2, PE), fluorescence detected in the FI1 (green fluorescence) channel, the setting for green fluorescence detection is typically not readjusted after the acquisition of the MESF bead data because the range of fluorescence in test cells is typically known.

Following hybridisation of nucleated human blood cells mixed with bovine thymocytes, three populations of cells can typically be distinguished based on two parameters: forward light scatter, which provides a measure of cell size, and LDS 751 fluorescence, which provides a measure of DNA content or accessibility. The three populations are bovine thymocytes (R1), which are dimly labeled with LDS 751, human lymphocyte (R2), which has intermediate forward scatter and LDS 751, and granulocyte (R3), which are mostly brightly stained with LDS 751.

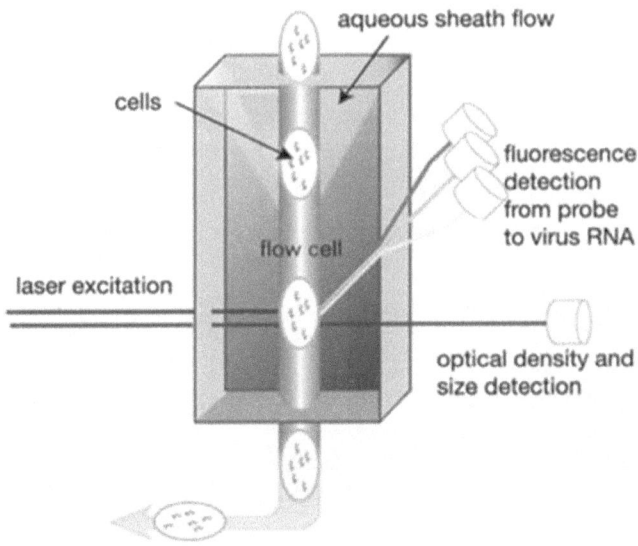

aqueous sheath flow

cells

fluorescence
detection
from probe
to virus RNA

flow cell

laser excitation

optical density and
size detection

Figure 10.1: A flow cytometry technique

Lymphocytes are further separated from granulocytes by combining the gates with the gates set in dot plots of side scatter versus forward light scatter. Cells within the lymphocyte gate (R2 + R4) are further subdivided on the basis of antibody labeling with Cy5 or Allophycocyanin and PE. The PE signal is derived directly from labeled CD20 antibody or indirectly from biotinylated CD57 antibodies, followed by Streptavidin-PE. After the identification of these various gates, the following populations can be distinguished: bovine control cells (R1 + R4), lymphocytes (R2 + R4), granulocytes (R3 + R5), PE-positive lymphocytes (R2 + R4 + R6), PE-negative lymphocytes, CD45RA-positive lymphocytes (R2 + R4 + R7), and PE-negative, CD45RA-negative lymphocytes (R2 + R4 + R8). Each of these populations of cells is assessed for their autofluorescence by analysing green fluorescence in tubes that were processed as for flow FISH in the absence of a PNA probe and blue peaks and for the fluorescence obtained with the telomere PNA probe and red peaks. At least ten thousand events of human cells are typically acquired for each analysis. Figure 2 is a typical image of a flow cytometry analysis.

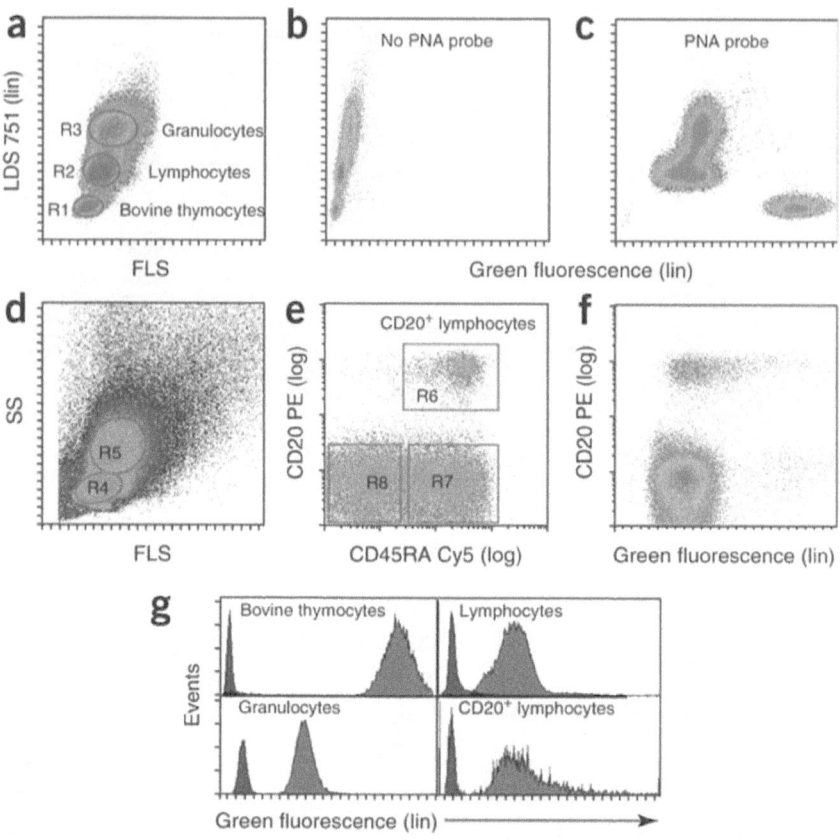

Figure 10. 2: Example of flow FISH data analysis based on nucleated blood cells from a normal human donor. For each nucleated blood sample, two samples are analysed: one in which the cells were hybridised to the peptide nucleic acid (PNA) probe (**c**) and one that was treated identically but without the PNA probe (**b**). The latter is required to measure the level of autofluorescence in cells of interest and to enable telomere length to be calculated from specific PNA hybridisation (**g**). Cells are counterstained with non-saturating concentrations of the DNA dye LDS751 and various antibodies (CD45RA–Cy5 and CD20–phycoerythrin (PE) in this case) before the acquisition of listmode data. The first step in the subsequent analysis is to identify cells using forward light and side scatter in a bivariate dot plot (**d**). Within gates R4 and R5, three cell populations can be observed in a bivariate plot of forward light scatter signal versus LDS751 fluorescence (**a**). The mild formaldehyde fixation of the bovine thymocytes

limits their staining by LDS751, which is useful to distinguish these small cells from human lymphocytes with largely overlapping forward and side light scatter properties. Granulocytes are labeled more brightly by LDS751 and can be distinguished from lymphocytes. The green fluorescence of cells gated as in (a) hybridised in the presence or absence of fluorescein-labeled PNA is shown relative to LDS751 fluorescence in the contour plots shown in (c) and (b), respectively. By combining the gates shown in (a) and (d), fluorescence histograms (g) of the indicated cell populations are obtained, which are used for subsequent calculations of telomere length. Antibodies specific for CD45RA and CD20 cells are used (e) to perform telomere length analysis of specific populations within the lymphocyte gate (R2 + R4). Note that the fluorescence histogram of granulocytes is more symmetrical than that of lymphocytes and that cells with relatively long telomeres are readily identified among CD20$^+$ B lymphocytes. (Source: Baelocher et al., 2006)

Most of the drug targets used for the treatment of human diseases are found in the human genome sequences; however, in infectious diseases, the drug targets are represented by the genome of both the host and pathogen. Therefore, a systematic functional genomics approach is needed to elucidate the changes that take place in the transcriptomes and proteomes of both the host and pathogen. Proteomic has been used in studying bacterial and fungal pathogens, but the technique is still in development. Flow cytometry (DNA microarrays) is an ideal technique for studying changes in the transcriptome of both host and oncoviruses. High-density DNA arrays are generated by spotting DNA fragments which are derived by PCR or synthetic oligonucleotide onto a solid surface such as glass. Oligonucleotide can also be generated on glass wafers using photolithography. Deposition of thousands of probes on a solid support surface allows the simultaneous monitoring of the expression levels of the corresponding mRNAs, which is isolated from various sources. DNA arrays are essential in analysing host-pathogen interaction. LNA-flow FISH can be used as a fast and easy way to screen for compounds with antiviral activity and could be used to monitor infections in blood for vaccine therapy and development. This chapter will review some of the data on DNA arrays, with emphasis on Merkel cell polyomavirus.

Polymaviruses (PyV) are naked, circular, ds-DNA viruses that infect birds and mammals (a fish-associated polyomavirus has been described). The genome of most polymaviruses is about 5000bp, which encodes regulatory and structural proteins. The major regulatory proteins are the large tumour antigen (LT-ag) and the small tumour antigen (ST-ag) that share conserved functional domain, made of binding motifs for the tumour suppressor pRB and p53 as well as for protein phosphatase 2A, respectively. Two structural proteins (VP1 and VP2) form the capsid. The regulatory proteins are expressed early during infection and play a part in viral replication and transcription. The structural proteins are expressed later in the infectious cycle. Most of the polymaviruses encode additional regulatory and structural proteins such as ALTO, VP3, VP4, and agnoprotein. Eleven novel human PyV have been described: KIPyV, WUPyV, Merkel cell PyV (MCPyV), HPyV6, HPyV7, Trichodysplasia spinulosa-associated PyV (TSPyV), HPyV9, HPyV10 (and isolates of MW and MX), STLPyV, PHyV12, and NJPyV-2013. MCPyV is associated with cancer in its natural host. About 80 per cent of Merkel cell carcinoma (MCC) tumours are positive for MCPyV genome, which is integrated and encodes a truncated form of LT-ag. Specific inhibitors against HPyV such as MCPyV are still lacking, while the development of a vaccine is still in the preliminary stage. Therefore, just like with other oncoviruses, identification of a unique technique that will aid in the acceleration of the development of anti-oncoviral drugs will be of immense help in the management of viral-associated cancers. Microarray techniques have been used in a number of studies which helped in our understanding of the pathogenesis of some viruses. In RSV infection, for example, microarray was used to identify genes involved in the pathways of neuroactive ligand-receptor interaction, p53 signaling, ubiquitin-mediated proteolysis, Jak-STAT signaling, cytokine-cytokine receptor interaction, hematopoitic cell lineage, cell cycle, and apoptosis. Respiratory disease biomarkers like APG2, SCNN1G, EPB41LAB, CSF1, PTEN, TUBB1, and ESR2 were detected with microarray. Flow cytometry in conjunction with cell culture has been used to analyse ganciclovir and foscarnet susceptibility for human cytomegalovirus clinical isolates.

There are two approaches to identify drug molecules: 1.Targeting key viral function, and 2. Using other broad-spectrum drugs that were originally designed for the treatment of another infection. Although there are no data on drugs developed utilising the technique, a lot of research has been undertaken using this technique to either identify new drug targets or analyse existing drug. In a study, Jiang et al. used FAC to identify a novel anticancer drug which targets the Wnt/B-catenin pathway. Sharma et al. studied the effect of artesunate on polyomavirus BKV. Using flow cytometry to analysis cell cycle, they found that artesunate decreased the release of infectious progeny in a concentration-dependent manner. They therefore concluded that artesunate inhibits BKV replication in RPTECS.

Measurement of binding of biotherapeutic or its cellular target, receptor occupancy, is now becoming an important tool in the development of biologically based therapeutic agents. RO assay by flow cytometry describes the quantitative and qualitative assessment of the binding of a therapeutic agent to its cell surface target. Taking into consideration that most oncoviral agents utilise receptor, this can be an ideal technique for drug development in oncoviral infection. Such assay can be as simple as measuring the number of cell surface receptors bound by anti-receptor therapeutic agents, or it can be designed to address more complicated issues such as internalisation, fusion, shedding, or assembly events. Data generated from RO assays can also be used to model pharmacodynamic (PD) markers during drug development as well as during preclinical and clinical trials. Drug development PD biomarkers are mostly used as surrogate markers to monitor changes in the biological effects of drugs. Data collected from assessment of these markers can be used in conjunction with pharmacokinetic (PK) data to aid in the efficient and effective design of clinical trials. Flow cytometry is a unique technique which can be used for oncoviral drug development. However, a few challenges will be encountered when using this technique, including maintaining the stability of the target-bound therapeutic agent and the stability of the target receptor. Also, reagent selection will be an issue, as reagents need to be evaluated for their potential to compete with the therapeutic agent and bind with comparable affinity.

REFERENCES

Allander T, et al (2007), Identification of a third human polyomavirus. J Virol 81: 4130–4136.

Baerlocher GM, et al (2006), Flow cytometry and FISH to measure the average length of telomeres (flow FISH). Nature Protocols 1: 2365–2376.

Buck CB, et al (2012), Complete genome sequence of a tenth polyomavirus. J Virol 36: 10887.

Chee M, et al (1996), Accessing genetic information with high-density DNA arrays. Science 274:610–614.

DeRisi JL, et al (1997), Exploring the metabolic and genetic control of gene expression on a genomic scale. Science 278:680–686.

Feng H, et al (2008), Clonal integration of a polyomavirus in human Merkel cell carcinoma. Science 319: 1096–1100.

Gaynor A, et al (2007), Identification of a novel polyomavirus from patients with acute respiratory tract infection. PLos Pathog 3: 595–604.

Green CL, et al (2015), Recommendation for the development of flow-cytometry-based receptor occupancy assays. Cytometry B Clin Cytom doi: 10.1002/cyto.b.21339.

Hedley DW, et al (1985), Application of DNA flow Cytometry to paraffin-embedded archival material for the study of aneuploidy and its clinical significance. Cytometry 6: 327–333.

Ideker T, et al (2001), Integrated genomics and proteomic analyses of a systematically perturbed metabolic network. Science 292:929–934.

Jiang HL, et al (2015), Wnt/B-catenin signaling pathway in lung cancer cells is a potential target for the development of novel anticancer drugs. J BUON 20: 1094–1100.

Lim ES, et al (2013), Disxocery of STL polyomavirus, a polyomavirus of ancestral recombinant origin that encodes a unique T antigen by alternative splicing. Virology 436: 295–303.

Lipson SM, et al (1997), Antiviral susceptibility testing-flow cytometric analysis (AST-FAC) for the determination of cytomegalovirus drug resistance. Diagn Microbiology & Infectious Diseases 28: 123–129.

Lowe DB, et al (2005), In vitro Simian virus 40 large tumor antigen expressions correlates with differential immune response following DNA immunization. Virology 332: 28–37.

Johne R, Buck CB, et al (2011), Taxonomical development in the family polyaviridae. Arch Virol 156:1627–1634.

Korup S, et al (2013), Identification of a novel human polyomavirus in organs of the gastrointestinal tract. PLOS ONE 8: e58021.

Manger ID, Relman DA (2000), How the host "sees" pathogens: Global gene expression response to infection. Curr Opin Immunology 12: 215–218.

Meckes DG, et al (2013), Modulation of B-cell exosome proteins by gamma herpesvirus infection. PNAS USA 110: E2925–E2933.

Michelson AD (1996), Flow Cytometry: A clinical test of platelet function. J of America society of Heamatol 87: 4925–4936.

Mishra N, et al (2014), Identification of a novel polyomavirus in a pancreatic transplant recipient with retinal blindness and vascular myopathy. J Infect Dis 210: 1595–1599.

Moens U, et al (2015), The role of Merkel cell polyomavirus and other human polyomaviruses in emerging hallmarks of cancer. Viruses 7: 1871–1901.

Peretti A, et al (2015), Genome sequence of a fish-associated polyomavirus, Black sea Bass (*Centropristis striata*) polyomavirus I. Genome Announcement 3: e01476–14.

Sauvage V, et al (2011), Human polyomavirus related to the green monkey lymphotropic polyomavirus. Emerg Infect Dis 17: 1364–1370.

Schowalter RM, et al (2010), Merkel cell polyomavirus and two previously unknown polymaviruses are chronically shed from human skin. Cell host microbe 7:509–515.

Scuda N, et al (2011), A novel human polyomavirus closely related to the Africa green monkey-derived lymphotropic polyomavirus. J Virol 85: 4586–4590.

Sharma BN, et al (2014), Antiviral effects of artesunate on polyomavirus BK replication in primary human kidney cells. Antimicrob Agents Chemother 58: 279–289.

Stewart JJ, et al (2015), Role of receptor occupancy assays by flow cytometry in drug development. Cytometry B Clin Cytom, DOI: 10.1002/cyto.b. 21355

Svebrasse EA, et al (2012), Identification of MW polyomavirus, a novel polyomavirus in human stool. J Virol 86: 10321–10326.

Teissier E, et al (2011), Mechanism of inhibition of enveloped virus membrane fusion by the antiviral drug Arbidol. PLoS One 6: e15874.

Wyant T, et al (2015), Development and validation of receptor occupancy pharmacodynamic assays used in the clinical development of the monoclonal antibody Vedolizumab. Cytometry B Clin Cytom doi: 10.1002/cyto.b.21286.

Van der Meijden E, et al (2010), Discovery of a new human polyomavirus associated with trichodysplasia spinulose In immunocompromised patients. PLoS Pathog 6: e1001024.

Velders MP, et al (2001), Human T cell response to endogenously presented HLA-A*0201 restricted peptides of simian virus 40 large T antigen. J Cell Biochem 82: 155–162.

Venter et al (2001), The sequence of the human genome. Science 291:1304–1351.

Vlastos A, et al. (2003), VPI pseudocapsids, but not a glutathione-S-transferase VPI fusion protein; prevent polyomavirus infection in a T-cell immune deficient experimental mouse model. J Med Virol 70: 293–300.

Yu G, et al. (2012), Discovery of a novel polyomavirus in acute diarrheal samples from children. PLOS ONE 7: e49449.

ABOUT THE AUTHOR

 **Abubakar Yaro** is a virologist and infectious diseases scientist who is interested in elucidating the mechanisms whereby dys- and upregulation of immune system are involved in the pathogenesis of viral and other infectious diseases. He is a faculty at AHRO Institute of Health Sciences & Research as well as the Founder and editor-in-chief of *Annals of Tropical Medicine & Public Health*.